Eco-design for Buildings and Neighbourhoods

Eco-design for Buildings and Neighbourhoods

Bruno Peuportier
Centre for Energy Efficiency of Systems of MINES Paristech

CRC Press
Taylor & Francis Group
Boca Raton London New York

CRC Press is an imprint of the
Taylor & Francis Group, an **informa** business

A BALKEMA BOOK

Photographs on cover:

Plus Energy houses in Freiburg (Germany)
Plus energy school in Limeil-Brevannes (France)
Photovoltaic system integrated in an existing building, Freiburg (Germany)

CRC Press
Taylor & Francis Group
6000 Broken Sound Parkway NW, Suite 300
Boca Raton, FL 33487-2742

First issued in paperback 2019

© 2016 by Taylor & Francis Group, LLC
CRC Press is an imprint of Taylor & Francis Group, an Informa business

No claim to original U.S. Government works

ISBN-13:978-1-138-02795-4(hbk)
ISBN-13:978-0-367-37729-8(pbk)

Typeset by MPS Limited, Chennai, India

Library of Congress Cataloging-in-Publication Data
Applied for

**Visit the Taylor & Francis Web site at
http://www.taylorandfrancis.com**

**and the CRC Press Web site at
http://www.crcpress.com**

Table of contents

Foreword

For over twenty years, designers, architects and engineers have been trying to convince their partners, contractors and local authorities to urgently rethink practices, methods and propositions to minimize the impact that buildings have on the environment.

That issue is now crucial.

The challenge is clearly technical, but also, and above all, social, financial and cultural.

It brings into question our behaviour, lifestyle and the way we relate to each other. The challenge is particularly linked to usage, and therefore to "Others", those that we so often forget: the people who will be living in the building, those who will maintain it, visit it, even those who will never set foot in it, but whose living environment will be changed by the new construction, from near or far.

New technical expertise allows us to build while taking care of the environment, bringing with it a simple reminder of the need for common sense, long-term planning, usage and comfort, as well as a source for innovation and high technology.

This new knowledge raises new issues in contemporary architecture. Let us hope it will encourage us to change the way we think about architecture and express it in our towns and elsewhere.

Françoise-Hélène Jourda

About the author

Bruno Peuportier holds an engineering diploma from the Ecole Centrale de Paris and a PhD from Université Paris VI. He is presently senior scientist at the Centre for Energy efficiency of systems of MINES Paristech. He has developed software tools for green design: COMFIE (thermal simulation) and EQUER (life cycle assessment). He has carried out several demonstration projects regarding the construction or renovation of social housing, office buildings and schools, and has coordinated a number of European projects in these fields.

Acknowledgements

This book owes much to those pioneers who, at a time when environmental quality was not a topical issue, realized the importance of saving resources and preserving the environment. Examples include the brochure "Energie solaire et architecture" written by Frédéric Nicolas, Jean-Pierre Traisnel and Marc Vaye in 1974, the solar wall designs invented by Félix Trombe and Jacques Michel in 1971, the solar housing competition organized by Alain Liébard and his team, and the initiatives of the Comité de Liaison Energies Renouvelables, all of which have made significant contributions to sustainable development in the building domain.

Additional studies on managing energy demand include the work of Olivier Sidler and the book "La maison des néga-watts" by Thierry Salomon and Stéphane Bedel. I received invaluable advice on bioclimatic architecture from Pierre Diaz Pedregal, and Renaud Mikolasek's work on developing user-friendly interfaces has helped disseminate our calculation tools among professionals. This book has also benefitted from input from my colleagues at the Ecole des Mines de Paris, in particular Isabelle Blanc Sommereux, Bernd Polster, Alain Guiavarch, Emil Popovici, Stéphane Thiers, Maxime Trocmé and Bruno Filliard.

Our knowledge of the links between buildings and health owes much to the publications of Doctors Suzanne and Pierre Déoux. Some enthusiastic people have launched themselves, often in difficult conditions, into building or spreading environmentally friendly techniques, such as solar collectors, renewable building materials, and water-saving equipment. My thanks go to the contractors, designers, architects, engineering firms and companies that have experimented with these innovations, as well as to the institutions that have supported their efforts.

The links between pollution emissions and environmental impacts have been much more widely studied in the Netherlands than in France, and their results have been used in this book. Lastly, I would like to thank my Swiss and German colleagues, in particular the developers of the Ecoinvent database and Niklaus Kohler, who established and disseminated data on environmental impacts of the main products and procedures that have developed our knowledge of aspects still not fully understood.

Taking inspiration from these forerunners, this book invites you to help protect our environment.

Introduction

I THE ENVIRONMENTAL QUALITY OF BUILDINGS

The notion of quality is complex in the building sector because of the variety of functions that installations need to fulfil[1], such as providing a space suitable for carrying out diverse activities, with given comfort conditions, protecting possessions and people, managing relationships with the outside, projecting an image, etc. The quality criteria concern different technical areas (e.g. structural mechanics, acoustics, lighting, heat transfer) as well as more subjective areas (e.g. aesthetics, quality of life, comfort).

In this book we introduce an additional area: environmental quality, which is linked to the quality of interior atmosphere and the reduction of buildings' environmental impacts. The aim is to disseminate knowledge resulting from research as a complement to other approaches, particularly architectural ones. Sustainable design, which involves taking environmental aspects into account, bolsters the dialogue between architecture and engineering. Other areas, like medicine, add into the equation. The first chapter defines the main environmental issues on different scales: global, regional and local, down to the indoor conditions.

The second chapter introduces indicators that correspond to the environmental issues covered in chapter 1. These indicators allow us to connect decisions with their potential impact or damage, thus making it easier to include environmental aspects in the decision-making progress. We present other indicators used to assess a town's environmental state. These developments shed light on the contribution that buildings make to the urban environmental balance, and this type of analysis can also orientate the choice of a site for a new construction.

Environmental indicators are integrated into a much bigger structure that includes economic, social and cultural aspects in a "sustainable" development approach (*durable* in French, *sostenible* in Spanish, *nachhaltig* in German). "Sustainable development is development that meets the needs of the present without compromising the ability of future generations to meet their own needs[2]". The Brundtland

---

[1]Christophe Gobin, Analyse fonctionnelle et construction, Techniques de l'Ingénieur, Dossier C3-052, November 2003
[2]Brundtland report, *Our Common Future*, report by the World Commission on Environment and Development, 1987

Report's definition stipulates that sustainable development "contains within it two key concepts:

- the concept of 'needs', in particular the essential needs of the world's poor, to which overriding priority should be given; and
- the idea of limitations imposed by the state of technology and social organization on the environment's ability to meet present and future needs."

The third chapter proposes several methods and tools for professional use. The life cycle assessment method, used to evaluate a product's environmental quality, is presented in its application in the building sector, including the connection between this method and energy calculations. The COMFIE software programme developed by the Ecole des Mines is considered because it can precisely evaluate both heating requirements and the degree of thermal comfort (dynamic, multi-zone simulation).

Chapter 4 sets out several technical solutions that can be used in the building sector to reduce environmental impacts. They concern the materials and components used, and more particularly flow management during the operation phase, which is crucial in the total emissions balance. We therefore present technologies for managing energy, water and waste flows.

Several examples of buildings are given in chapter 5 as an illustration. The end of this chapter includes general recommendations for dealing with the different project phases, from feasibility studies to end of life.

This book is aimed at a varied readership: clients, architects, engineering firms, contractors, manufacturers of materials and components, property managers, and urban planners. Depending on their previous knowledge and particular interests, readers can use the contents to guide them. The principles of sustainable design presented here should not be taken as a strict method, but an approach to complement the aspirations of contractors and the creativity of architects.

Chapter 1

Environmental impacts

Buildings create an intermediate space between occupants and the outdoor environment, with a role of offering a space suitable for the intended activities (e.g. housing, professional work), while fitting into a site. We can break this down into:

- The "indoor environment", which constitutes an environment for a building's occupants. This "built environment" must fulfil a number of quality requirements (e.g. functionality of spaces, hygrothermal, visual, acoustic and olfactory comfort, health safety and quality of life);
- The outdoor environment, from the building's surroundings, the immediate area, and the region (with spatial scales that may vary according to the administrative zoning and an analysis of environmental problems), through to global scale. The aim is to minimize impacts on these different scales (protection of climate, fauna and flora, resources, health and landscape);
- Relations between the indoors and outdoors, which also need to fulfil certain requirements: movement of people and goods, protection, aesthetic quality of the envelope, use of "natural" flows (solar energy, rain water), connection to water, energy and transport networks, waste management.

Added to these various spatial scales is the time scale. Buildings generally last for many years, and performance should therefore be long lasting. The issue of "sustainable" development also means taking a long-term view for the benefit of future generations (e.g. climate protection, long-term waste, biodiversity, genetic heritage, etc.).

Within this vast framework, a number of environmental issues are presented below, with a focus on aspects generally still rarely addressed in the building sector.

1.1 GLOBAL SCALE

From 1950 to 2013, the world's population expanded from 2.5 to 7.2 billion, and the number of towns with over 8 million inhabitants rose from 2 to 42[3]. One in five people in the world do not have access to drinking water. CO_2 emissions went up by

---

[3] http://www.citypopulation.de/world/Agglomerations.html, accessed 11/09/2013

70% from 1970 to 2004, approaching 50 billion tonnes of CO_2 equivalent per year[4]. Primary forests have lost 30% to 50% of their surface area, while 10 million hectares of tropical forest (i.e. almost a fifth of the area of France) and 1,000 species of plants and animals disappear each year.

Global considerations are the most difficult to take into account in this field for several reasons:

- Individuals play a miniscule role in the global balance sheet, which reduces motivation for making decisions in the general interest;
- Respecting the planet is not generally translated in economic terms, nor in an immediately perceptible way (reducing your CO_2 emissions has no direct impact on your own health and quality of life).

However, an ostensibly small modification to the global balance sheet can have considerable consequences. For example, the concentration of oxygen in the atmosphere is 21%. If it dropped to 16%, birds, reptiles and mammals would be asphyxiated, and if it rose to 25%, fires would rage, even in the rain forests[5].

1.1.1   The greenhouse effect

The atmosphere diffuses part of the sun's rays in all directions, but a high proportion is transmitted towards Earth, which heats up and radiates in turn. The range of a ray's wavelength depends on the temperature of the emitting body. Thus, the sun, whose surface temperature is around 5,800 K, emits visible radiation, whereas the Earth, whose average temperature is 15°C, emits long-wave radiation (infra-red).

Some gases present in the atmosphere (CO_2, water vapour, methane, etc.) are more transparent to the sun's radiation than to the Earth's infrared radiation, thus producing the greenhouse effect: the radiation is partly trapped, which provokes heating. The Earth emits around 390 W/m² at the surface, whereas the radiation emitted towards space is 240 W/m². The difference is called "radiative forcing". A doubling of the concentration of CO_2, compared to pre-industrial levels, could increase this value by 4 W/m², which would cause a temperature rise.

The greenhouse effect exists naturally, and without it the average temperature of the Earth would have been −18°C instead of about +15°C for the last 10,000 years (only +10°C 20,000 years ago during the ice age). It is its increase that brings problems. Human activity has resulted into a rise in greenhouse gas emissions: the CO_2 given off by using fossil fuels is responsible for 55% of the increase in greenhouse gases (during the 1980s), of which 25% CFCs (chlorofluorocarbons) and 15% methane, with nitrous oxide (N_2O), SF_6 and the ozone formed making up the remaining 5%[6]. The concentration of CO_2 in the atmosphere rose from 280 ppm (parts per million) before the industrial era to 379 in 2005. It is currently increasing by 0.5% per year

---

[4]Intergovernmental Panel on Climate Change (IPCC), Assessment Report No. 4, 2007, http://www.ipcc.ch
[5]*Systèmes solaires* magazine, No. 120, 1997
[6]Intergovernmental Panel on Climate Change (GIEC), mentioned by ADEME in their brochure *Effet de serre*, www.ademe.fr

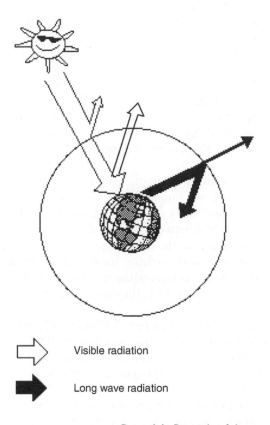

The atmosphere diffuses part of the sun's rays in all directions, but a high proportion is transmitted towards Earth, which heats up and radiates in turn. The range of a ray's wavelength depends on the temperature of the emitting body. Thus, the sun, whose surface temperature is around 5,800 K, emits visible radiation, whereas the Earth, whose average temperature is 15°C, emits long-wave radiation (infra-red).

⇨ Visible radiation

⬛▶ Long wave radiation

Figure 1.1 Principle of the greenhouse effect.

(the equivalent of 5 to 6 gigatonnes) and, at this rhythm, is set to double by 2060. Around 1800, the average French inhabitant consumed 70 kg of carbon equivalent per year[7]. Today, that figure has risen to about 5 tonnes (which is however half of what an average North American uses).

Each French inhabitant emits around 8.7 tonnes of CO_2 equivalent per year, taking all activities together[8]. An average North American emits almost three times as much (emissions rose by around 11% in the USA, the country with the highest emissions, from 1990 to 2002) and an Indian almost seven times less (although this country's emissions rose by 57% from 1992 to 2002, and by 33% in China over the same period). In comparison, through breathing, each individual releases about 16 litres of CO_2 per hour, or about 300 kg a year.

---

[7]Éric Labouze, *Bâtir avec l'environnement, enjeux écologiques et initiatives industrielles*, L'Entrepreneur, Paris, 1993
[8]CITEPA (Centre Interprofessionnel Technique d'Etudes sur la Pollution Atmosphérique), Inventaire des émissions de polluants atmosphériques en France – séries sectorielles et analyses étendues, February 2008, table p37, line "except UTCF" (use of land, land change, and forest)

Some international negotiations (e.g. Kyoto process) led to a commitment by a number of states to reduce their emissions from 2010–2012 compared to 1990. For example, the 15-state European Union was to reduce its emissions by 8% overall (with differences between countries, ranging from a 28% reduction for Luxembourg to a 27% rise for Portugal, with French emissions remaining stable). The Kyoto commitments represent an overall reduction of about 5% for developed countries. The European Union is taking a more proactive attitude, aiming for a reduction of 20% by 2020. Emissions in France have remained stable since 1990: 547 Mt of CO_2 equivalent in 2006, compared to a 1990 reference value of 566 Mt. From 1990 to 2006, emissions went up for CO_2 (+3.3%) and HFCs[9] (+1123%) but went down for CH_4 (−17.7%), N_2O (−29.5%) and SF_6 (−40.8%)[10].

Methane concentration doubled, rising from 1.8 ppm, and continues to rise by 0.9% per year. The lifespan of greenhouse gases ranges from 10 years for methane to 120 years for CO_2 (in fact, complex phenomena of CO_2 reabsorption in oceans and the biosphere point towards a time interval between 50 and 200 years). For this reason, the equivalents vary, partly depending on the optical qualities of gases and their lifespan in the atmosphere (cf. chapter 2), and partly on the duration considered in the analysis: over 100 years, 1 kg of methane is equivalent to 25 kg of CO_2; over 500 years, it is only worth 7 kg of CO_2 (1 kg of N_2O is the equivalent of 290 kg of CO_2).

Since 1850, the average global temperature has risen by 0.7 K and by 1 K in France according to Météo France. The concentration of CO_2, according to IPCC provisions[11], will rise by 1.4 to 5.8 K by 2100 based on evolution scenarios. The Earth's climatic zones would then move towards the poles (200 km per additional K) or change altitude (150 m per K), at a speed that plant species would not necessarily be able to adapt to. Maintaining the temperature rise between 1.5 and 4 K would entail restricting CO_2 concentration to 450 ppm, which corresponds to halving the 1990 level of global emissions by 2050.

Simulations carried out with "Arpège-Climat étiré"[12], which suppose a doubling of CO_2 concentration, evaluate a temperature rise in France of between 1 and 2 K in winter and over 2 K in summer, with wetter winters and dryer summers. The Gulf Stream, an ocean current that warms the coasts of Western Europe, might be perturbed or even disappear, bringing a drop in temperature and precipitations in Europe. Climate change also brings the development of cyclones in some regions, and summer droughts and heavy autumn rains in our regions. According to the European Environment Agency, 75% of the Swiss glaciers are likely to have disappeared by 2050[13].

Meteorological models cannot yet ascertain how global warming influences the frequency and strength of storms. Insurance and reinsurance companies, concerned

---

[9]Hydrofluorocarbons are fluids used in air conditioning systems and heat pumps
[10]www.citepa.org, under publications, then inventories
[11]Intergovernmental Panel on Climate Change, created in 1988 by the World Meteorological Organization and the United Nations Environment Programme, http://www.ipcc.ch
[12]Pierre Bessemoulin, *Changement de climat en France au XXIe siècle?* Pollution atmosphérique No. 165, January–March 2000
[13]http://www.eea.europa.eu/media/newsreleases/soer2005_pp-en

about the increase in natural disasters, suggest reinforcing application of the precautionary principle. The storm that hit France at the end of 1999 caused the deaths of 80 people. Damage was evaluated at 40 billion francs (over 6 billion euro), and 300 million trees were knocked down, or the equivalent of 100 million m³ of wood, 25% more than the average annual production. The floods in Germany in 2002 caused 27 deaths and 15 billion euro of damage.

If all the continental glaciers melted (in particular in Greenland and especially Antarctica), the sea level would rise by 80 metres. The Sarennes Glacier in the Grandes Rousses mountains is monitored by CEMAGREF. Melting over one century is 80% in mass, with maximum shrinkage of 3 metres per year[14]. The report by Senator Marcel Deneux[15] indicates that for a rise in temperature of 1.8°C, snow would diminish by 20% to 25% at 1500 m altitude in the northern Alps, and up to 45% in the Pyrenees. The snow period would be reduced by 30 to 50 days a year.

Figure 1.2 Evolution of Bossons Glacier (shrinkage of around 500 m from 1980 to 1995), source: Ecole des Houches.

The thickness of ice on the Arctic Ocean has diminished by 42%, and the North Pole is now in water in summertime[16]. Greenland loses 51 billion m³ of water each year, the equivalent of the flow of the Nile. From 1950 to 1997, the surface area of the Antarctic ice fields diminished by 7000 km², with a temperature rise of 2.5°C in the region.

According to models, the sea level should rise by +66 cm ± 57 cm by 2100, threatening 3% of land (500 million hectares) with floods or soil salinity. According to a study by the World Bank, a rise of one metre would submerge half of the rice paddies in Bangladesh. The islands of Tuvalu, the 189th state to join the United Nations, are victim to rising Pacific waters. Since creating dykes round the islands would be too costly (and in any case, the water filters up through the coral reef), the 11,000 inhabitants have been offered immigration to New Zealand. The highest point of Tuvalu is

---

[14]*Energie Plus* No. 244, 15 April 2000
[15]Marcel Deneux, Evaluation de l'ampleur des changements climatiques, de leurs causes et de leur impact prévisible sur la géographie de la France à l'horizon 2025, 2050, 2100, Parliamentary Office for Evaluation of Scientific and Technological Options, July 2002
[16]Lester R. Brown, *Climate Change Has World Skating On Thin Ice*, 29 August 2000

only 4.5 m above sea level[17]. France, with almost 7000 km of coastline, would also be affected by this phenomenon.

Economic impacts concern agriculture[18] (quantity and quality of products, need for irrigation, phytosanitary products, etc.), as well as water resources and health[19] (effect of heat waves, increase in tropospheric ozone concentration, development of infectious diseases like malaria). Given the uncertainty of the consequences of the greenhouse effect, the financial cost of these impacts is very difficult to assess. Some suggest figures of around USD 500 per tonne of CO_2[20], which corresponds to around 0.5 euro/electric kWh and 0.11 euro/thermal kWh (primary energy: gas). The greenhouse effect would also reduce the need for heating buildings (around 10% for a temperature rise of 1 K), but would involve using more air conditioning. However, even though current knowledge does not allow us to anticipate consequences with certainty, experts urge for caution and reducing emissions[21].

Solutions to reduce greenhouse gases[22] include not just saving energy, but the type of fuel used. For the same quantity of energy produced, the quantity of CO_2 emitted is shown in table 1.1 for different types of fuel, taking hard coal as a reference. However, the whole chain needs to be taken into account, from extraction to utilization, and including distribution. Natural gas, for example, involves distribution losses of up to 2.5% (in gas pipelines, etc.). Despite the high emissions resulting from burning coal, the global consumption of this fuel continues to rise, by 9% in 2006.

Table 1.1 CO_2 emissions for different fuels.

Fuel	Lignite	Hard coal	Oil	Natural gas	Wood
CO_2 emitted (ratio/reference to equal energy produced)	121	100 (reference)	88	58	0*

(*)CO_2 emitted is compensated by CO_2 absorbed by the forest

Renewable energies also help protect the climate, since less energy is required for manufacturing the systems than for production (emissions balance sheets are also favourable, cf. chapter 3).

At global level, the carbon stored in wood probably represents 500 billion tonnes, which is about the same as that stored in the atmosphere[23]. Growing trees capture carbon more efficiently. Therefore using wood and logging favour carbon storage,

---

[17]Politis, 29 November 2001

[18]Richard Delecolle, Pierre-Alain Jayet and Jean-François Soussana, *Agriculture et réchauffement climatique, quelques éléments de réflexion*, La jaune et la rouge, 2000

[19]Jean-Pierre Besancenot, *Le réchauffement climatique et la santé*, La jaune et la rouge, 2000

[20]Olav Hohmeyer and Michael Gärtner, *The cost of climate change: a rough estimate of orders of magnitude*, The Yearbook of Renewable Energies, EUROSOLAR/EEC, 1994

[21]Yves Martin, *L'effet de serre*, Report by the Académie des Sciences, No. 31, November 1994

[22]*L'effet de serre et la Communauté européenne*, Revue générale de thermique, No. 329, May 1989

[23]Éric Labouze, *Bâtir avec l'environnement, enjeux écologiques et initiatives industrielles*, Editions de l'Entrepreneur, Paris, 1993

and the Rio Conference recommended their expansion, in particular in the building industry[24]. While fuel wood is becoming rare in some southern countries, it appears to be underused in developed countries, and the Rio Conference also recommended that these countries should increase their use of fuel wood.

Photosynthesis is a complex set of chemical reactions, but put simply, cellulosic matter is produced from CO_2 and water. The corresponding mass balance indicates that to produce 1 kg of cellulose requires around 1.5 kg of CO_2. It takes several kg of tree to obtain 1 kg of construction wood. The wood must then be cut, transported, dried and sawn, all of which creates emissions. According to the Oekoinventare database[25], 1.85 kg of CO_2 is stored to produce 1 kg of planks or beams. Incineration re-emits CO_2 among other things, but the heat produced can be used. Disposing of the wood can generate methane, but this gas can be collected. The balance of wood is therefore of interest, and the Air Law of 1996 recommends that it should be used more in buildings. But the 2005 implementing decree fixes the limit at 2 dm^3/m^2 of built surface area, whereas the average current use of wood is 30 dm^3/m^2. The annual flow of CO_2 harnessed by the French forest is evaluated as 60 Mt, i.e. 12% of national emissions (except during the 1999 storms, which temporarily cancelled out this effect)[26].

The manufacture of CFCs[27] is now forbidden, but it is possible that residual clandestine manufacture continues in some countries, and some substitutes (HFC, and HCFC – banned from 2020 in industrialized countries and from 2032 in developing countries) also contribute significantly to radiative forcing. Methane emissions could be reduced (by improved extraction procedures, transport and use of natural gas, recuperation or incineration at landfills, better management of livestock, changes in rice farming and lagoon management). Reducing the use of nitrogenous fertilizer in farming would avoid some emissions of nitrous oxide.

The cost of reducing emissions varies from several cents to over USD 400 per tonne according to studies[28]. The economic optimum is when the marginal cost of reducing emissions equals the marginal cost of the avoided impact. Given the uncertainty of these two terms, achieving such an optimization is difficult. It is however possible to encourage 'no regret' strategies, i.e. no supplementary cost. For example, energy-saving measures in buildings, wherein investments are profitable over their lifespan. Table 1.2 gives an overview of the estimated cost of various types of prevention listed at European level[29].

These costs need to be linked to the price of a tonne of CO_2 on the European market, which fluctuates from 6 to 20 euro, where the penalty for emissions without available authorization was 40 euro from 2005–2007 and 100 euro from 2008–2012.

---

[24]Christian Barthod, Ministry for Agriculture and Fisheries, *Chronique des négociations internationales récentes*, Aménagement et Nature magazine, No. 115, autumn 94
[25]Frischknecht R., et al., *Oekoinventare von Energiesystemen*, 3. Auflage, ETH Zürich/PSI Villigen, 1996
[26]Institut Français de l'Environnement, Les données de l'environnement No. 105, August 2005
[27]CFC: chlorofluorocarbon, HCFC: hydrochlorofluorocarbon, fluids used for thermodynamic cycles, e.g. air conditioning systems and heat pumps
[28]Jean-Charles Hourcade, *Traitement de l'innovation et évaluation des coûts à long terme de la réduction des émissions de CO_2*, Revue de l'Energie, No. 427, January 1991
[29]European Commission, *Economic Evaluation of Sectorial Emission Reduction Objectives for Climate Change*, 2001

Table 1.2 Cost of diverse measures for reducing CO_2 emissions and potential for europe (EU15).

Cost (1)	Prevention technique	Potential (2)
<0	Gas combined cycles and refineries	513
	Reducing industrial emissions	347
	Thermal insulation on existing buildings	81
	Recycling solid biomass	33
	Improving management of heating/air con.	42
	Managing electricity demand	36
	Improving goods vehicles	19
	Reducing emissions in farming	14
<10 euro	Reducing industrial emissions	150
	Better insulated windows in redevelopments	80
<20 euro	Using wood for heating	64
	Managing waste better	26
	Improving car design	24
	Improving transportation of goods	19
	Industrial improvements	8
<30 euro	Industrial improvements	35
	Onshore wind power	30
	Wood cogeneration	29
	Landfill replaced by incineration	23
	Condensing boilers	15
<50 euro	Hydroelectricity	17
	Industrial improvements	12
	Reducing emissions in farming	5
<100 euro	Cogeneration and CO_2 separation	76
	Improving car design	29
	Improving mines	14
<150 euro	Offshore wind power	18
<300 euro	Cogeneration	40
	Biodiesel and ethanol	33
	Solar power	8
<500 euro	Improving car design	21
	Heat pumps	16

(1) per tonne of CO_2 equivalent avoided
(2) in millions of tonnes of CO_2 equivalent avoided

A number of studies focus on CO_2 storage. In certain conditions, this gas, once dissolved in water, forms carbonate, which is harnessed by shells and associated with calcium to form their shell. When the animal dies, it sinks and part of the shell dissolves, while the other part forms sediment on the seabed. After several million years, the layers formed can be packed several metres thick to form limestone. Thus, some stones used for building constitute CO_2 storage.

According to some studies, CO_2 injected into the oceans at a depth of 100 to 500 metres could be stored for 50 years, since exchanges between the different ocean layers are very slow. The estimate is several centuries for injections at 1000 metres (CO_2 is liquid at the pressure that corresponds to this depth). From 3000 metres, the pressure makes the CO_2 heavier than water. However, the consequences of the acidification produced are as yet unknown. Activation of corals or phytoplankton are sometimes envisaged, but a great deal of uncertainty remains regarding the efficiency of this kind of procedure and potential side effects.

Capturing CO_2 in combustion installations (e.g. thermal power stations) and storing it in oil wells or disused mines brings risks for neighbouring inhabitants, since any CO_2 leakage could provoke asphyxia.

1.1.2 Destruction of the ozone layer

The atmosphere is divided into five zones: the troposphere (altitude <17 km), the stratosphere (from 17 to 50 km), the mesosphere (50 to 80 km), the ionosphere (80 to 600 km) and the exosphere. The troposphere, in which we live, is where air circulates and atmospheric phenomena occur (e.g. rain, clouds and cyclones). It contains almost all of the water vapour and most of the greenhouse gases. The ozone layer is located more in the stratosphere, which is where meteorites burn up. The temperature is very low in the mesosphere ($-100°C$), then rises again in the ionosphere, where X and gamma rays coming from the sun are stopped, and then reaches 1200°C in the exosphere.

The ozone layer is located at an altitude of between 12 km and 45 km, with a maximum concentration at 25 km. It filters almost all of the B-type ultraviolet rays, which attack the skin (cancer), eyes (cataracts) and the immune system (UVBs also upset the growth of some plants, destroy marine phytoplankton and accelerate the ageing of some materials[30]).

The ozone layer undergoes "natural" seasonal variations: its thickness above the South Pole diminishes during the polar nights (June to September). This is due to the formation of stratospheric clouds that capture nitrogen dioxide and prevent it from trapping chlorine atoms, which are responsible for diminishing the ozone. Clouds of volcanic aerosols, like those of Mount Pinatubo, can play the same role. The "hole" in the ozone layer actually corresponds to a surface where the ozone's thickness and concentration are below a certain threshold (220 Dobson units). According to the World Meteorological Organization, it exceeded 28 million km^2 in 2003.

This seasonal variation has got bigger and under our latitudes a reduction in thickness of around 5% on average is observed in winter time. CFCs are the main cause because they slowly rise up to the stratosphere, where they are destroyed by UVs and give off chlorine, which in turn provokes the catalytic destruction of the ozone layer. Other compounds involved in this phenomenon are halons (used in fire protection), which are responsible for 12% of the overall destruction of ozone. HCFCs can also play a role, and some products are regulated. HFCs, however, have no effect. The Vienna Convention (1985), the Montreal Protocol (16 September 1987), and the London Convention (1990) led 70 countries to ratify a total phase-out of CFCs by 2000. Their use has been banned in France since 1996. HCFCs will be banned in all industrialized countries in 2020 and in 2032 in developing countries.

23% of all CFCs consumed in France were used to manufacture insulation foam[31] (base of three-quarters polyurethane and one fifth extruded polystyrene). The average

---

[30] Serge Lambert et al., *Manuel environnement à l'usage des industriels*, AFNOR, 1994

[31] Éric Labouze, *Bâtir avec l'environnement, enjeux écologiques et initiatives industrielles*, Ed. de l'Entrepreneur, Paris, 1993

CFC content of this type of foam was about 10%, and the quantity of CFCs stored was estimated as 54,000 tonnes in 1989.

Prior to the ban, the cold sector and air conditioning, with over a million units loaded with an average 2 kg of fluid, represented between 10% and15% of French CFC consumption. A high proportion of gas (40% to 50%) was used for recharging operations. Current solutions include substituting other fluids and recuperating CFCs and HCFCs.

1.1.3   Depletion of resources

Some raw materials are becoming scarce, with varying cut-off points. Could we be in the process of reproducing on a global scale the scenario that probably occurred on Easter Island[32]? Oil is particularly affected, with reserves estimated at 1238[33] or 1300[34] billions of barrels, or around 40 times the current annual consumption level, but mercury, cadmium, tin, nickel, lead, zinc and copper are also becoming scarce.

Cut-off dates of several decades may seem far off, but the economic impacts of resource depletion can be felt well before with a rise in prices. The point occurs where consumption overtakes the volume of production, which sends prices shooting up. Production reaches a maximum point and then decreases (Hubbert's peak). Hubbert's peak has already been reached for European oil, and some claim that for the world it will be reached between 2010 and 2020[35]. Over 60% of oil reserves are in the Gulf States, which causes conflict in this zone. The main suppliers of oil to France in 2006 were Norway (13.4 Mt), Russia (9.8 Mt), Saudi Arabia (8.7 Mt), Kazakhstan (8.1 Mt), Iran (6.7 Mt), the United Kingdom (6.6 Mt) and Irak (4 Mt).

Proven global reserves of natural gas came to 177.4 thousands of billions of cubic metres at the end of 2006, which represents the equivalent of 60 years of production at its current level[36]. These reserves are mainly located in the Middle East (41%) and Russia (26%). In 2006, France imported 437 TWh of gas on long-term contracts, of which 34% from Norway, 22% from the Netherlands, 19% from Algeria, 19% from Russia, 6% from Egypt and 1% from Nigeria[37].

The annual production of natural uranium comes to 45,000 tonnes[38], 30% of which are in Canada, 21% in Australia, 10% in Niger (resulting in environmental and social issues, in particular close to the Arlit and Akokan mines exploited by AREVA),

---

[32]André Lebeau explains in his book *L'engrenage de la technique* (Gallimard, 2005) that the 15,000 people who lived on the island three centuries ago destroyed their forests, and that the resulting erosion made the soil unsuitable for farming, causing the end of their civilisation. Other authors, like C. and M. Orliac, attribute the island's desertification to climate conditions linked, for example, to the El Niño phenomenon (La flore disparue de l'île de Pâques, Les Nouvelles de l'archéologie No. 102)

[33]BP, Statistical Review of World Energy 2008, http://www.bp.com

[34]Annales des Mines, Réalités industrielles, August 2007

[35]ASPO, Association for the Study of Peak Oil and Gas, www.peakoil.net

[36]BP, Statistical Review of World Energy 2008, http://www.bp.com

[37]Annales des Mines, Réalités industrielles, August 2007

[38]French Ministry for Ecology, Sustainable Development and Energy, La lettre de la direction générale de l'énergie et des matières premières, No. 29, October 2007

and 6% in Kazakhstan, Uzbekistan and Russia[39]. Consumption totals 70,000 tonnes, feeding into around 430 civil nuclear reactors. Production is lower than consumption levels due to fuel recycling and the use of certain stocks. However, now that stocks resulting from dismantling nuclear weapons have run out, prices have risen and production has started up again: the price of a pound of uranium rose from 10 USD in 2001 to 125 USD in 2007.

Conventional reserves are estimated at 4.7 million tonnes[40] (i.e. 70 years' consumption) of which 24% are in Australia, 17% in Kazakhstan, 9% in Canada, 7% in the United States, 7% in South Africa, 6% in Brazil, 6% in Namibia, 5% in Niger, and 4% in Russia. Although the nuclear industry is frequently associated with energy independence, French production only represents 10% of the country's consumption and this volume is decreasing.

In primary energy terms, these reserves only represent 30% of oil reserves[41], and 3% of total fossil fuel reserves (including gas and coal). The use of non-conventional reserves would triple this potential, but at a much higher cost, depending on ore dilution: uranium content can be as high as 12% in some currently exploited deposits, but below a certain threshold, more energy is needed to extract the uranium than it could produce, which restricts the non-conventional potential.

Global coal reserves are estimated at 509 Gtoe (Giga tonnes of oil equivalent[42]), i.e. 230 years at current consumption levels. This fuel provides 1.5% of French electricity production[43], but is now totally imported from Australia, South Africa and the United States. The last French coal mines closed down during recent years: Forbach (1997), Vouters (2001), Merlebach (2003) and La Houve (2004).

Reserves of some tropical hardwoods, subject to badly managed logging, are also diminishing. The surface area covered by tropical forests has halved[44], with between 60,000 and 150,000 km^2 disappearing each year (i.e. 1% of the total forest area, and up to 6% in western Africa). Deforestation is mainly due to land clearing aimed at increasing crop areas. Logging only represents 6% of deforestation[45], and 90% can be put down to agricultural activities. 85% of global tropical hardwood production is used for firewood, 10% for local construction and only 5% is exported[46]. Global annual wood consumption was 3.5 billion m^3 in 1992[47].

Although France possesses abundant local resources, it is one of the leading importers of tropical hardwoods in Europe, at 1.3 million m^3 per year[48], 65% of which

---

[39]*L'état du monde*, Ed. La Découverte, Paris, 1996

[40]French Ministry for Ecology, Sustainable Development and Energy, La lettre de la direction générale de l'énergie et des matières premières, No. 29, October 2007

[41]Taking one kg of natural uranium produced to use 50,000 kWh of electricity, i.e. 130,000 kWh of primary energy, or 11 Toe, and one barrel (159 litres) to produce 1560 kWh i.e. 0.134 Toe

[42]1 tonne of oil equivalent corresponds to 11,600 kWh, 1 Giga tonne is equal to 1 billion tonnes

[43]http://www.rte-france.com/sites/default/files/bilan_electrique_2014.pdf, Réseau de Transport de l'Electricité, 2014

[44]R. Barbault, *Écologie générale*, Ed. Masson, Paris, 1990

[45]Study by the Kiel Institute for the World Economy, produced for Greenpeace

[46]Éric Labouze, *Bâtir avec l'environnement, enjeux écologiques et initiatives industrielles*, Ed. de l'Entrepreneur, Paris, 1993

[47]François Ramade, *Écologie appliquée*, Ediscience, Paris, 1995

[48]Que Choisir, *guide pratique des produits propres*, Nov./Dec. 1990

is used for construction. Some of this tropical hardwood is imported illegally (some say as much as 40%[49] to 50%[50]), for example from the Congo Basin (Democratic Republic of Congo, Gabon, etc.). The main species of tree concerned are okoume and meranti (used to make plywood, which represents over 50% of the tropical hardwood market), niangton, moabi, mengkulang and framire (carpentry), and iroko (parquets). These species grow in Africa, except for meranti, which is Asian. Brazil has started to export hardwoods. Oak and chestnut are potential substitutes, especially with the development of new treatments (e.g. impregnation of acetic acid and rapeseed oil derivatives) – chapter 4 looks at how to choose the most suitable species for different uses. Two million tonnes from old-growth forests in Canada and one million tonnes from Russia are also used in France for paper pulp, but 25% of the pulp produced from French wood comes from sawmill cut-offs and 75% from wood cut for clearing (pruning necessary for the forest's development). Paper consumption is rising constantly at a rate of 2% per year.

The Isoroy group, the main producer of French plywood, has created its own label "Eurokoumé", certifying that the material used has not upset the ecosystem[51] (Gabonese logging is restricted to 1% of the country's okoume growth). The FSC (Forest Stewardship Council) and PEFC (Programme for the Endorsement of Forest Certification) labels are presented in the section on biodiversity below.

In France, 64 million m³ of wood were harvested in 1999[52]. Since the 19th century, the surface area covered by French forests has doubled. In 2001, almost 16 million hectares were covered by forest, extending by around 40,000 hectares a year due to agricultural abandonment. Following the storms of 1999, exports doubled in 2000, but French forests returned to their initial volume in two years.

Plastic manufacture consumes 4% of oil production. The energy content of this material is usable at end of life, and some plastics can be recycled. The sector's contribution to depleting oil resources is therefore considered to be low.

The most widely present materials on our planet are (in descending order of proportion in the Earth's composition): oxygen (46%), silicon (27%), aluminium, iron, sodium, potassium and magnesium. Concrete is the second most consumed material after water, at 900 kg per person per year. In France, 420 million tonnes of aggregate (of which 85 million tonnes used in buildings) and 20 million tonnes of cement are consumed annually.

The mining law differentiates two types of raw material: the mine class (coal, oil, gas, uranium, metals, salt, potassium, phosphates), and the quarry class, which includes most building raw materials[53]. Landowners require a mining concession and planning permission, whereas quarry operators need to acquire land in line with the

---

[49]Fraudulently obtained concessions, invalid forestry licences, unpaid taxes, false customs declarations, etc. (The EU FLEGT timber regulation adopted end 2005 imposes a ban on wood that does not respect forestry regulations), Environnement Magazine No. 1648, June 2006
[50]Sustainable Building, No. 4, Netherlands, 2001
[51]System formed through the interaction of living beings (or biotic community) with their geological, pedological and atmospheric environment (biotope)
[52]Institut Français de l'Environnement, Les données de l'environnement No. 105, August 2005
[53]Jean Tassin, *L'accès aux ressources en matériaux de construction*, Energie et matières premières No. 10, 4th quarter 1999

deposit or obtain the owner's agreement, and respect planning regulations. In some cases, depending on a council decree following a public inquiry, the state may create "exclusive zones for quarry research and exploitation", known as "109 zones" so as to authorize the exploitation despite the owner's opinion. The prefect may also undertake a general interest project procedure.

Raw material consumption can be reduced by making components lighter (some lightweight concrete only uses 50% of the materials used by traditional construction, lightweight PVC can save 30%), improving performance (reducing the thickness of concrete structures, tensile strength steel consumes between 20% and 25% less materials), recycling (up to 60% for copper[54], 50% for steel used in buildings – but steel from demolitions is not always recycled – between 30% and 50% for aluminium, lead and zinc) and re-use. In 1992, 2000 tonnes of plastic were recycled in the building industry (flooring, tubes and cable troughs).

Energy consumption in France in 2013 totalled 165 Toe in final energy, of which 41% was oil, 21% gas, 19% nuclear (representing 77% of electricity production), 10% renewable energy (including hydroelectricity), and 4% coal[55]. A primary energy balance sheet allows us to assess the impact in terms of resource depletion. In fact, more energy is required at the start to produce one kWh of electricity than one kWh of heat. The equivalence factor between primary energy and final energy depends on the yield of the energy production chain, for example 3.5 kWh of primary are needed to produce 1 kWh of final nuclear electricity (taking into account extraction, transport and enrichment of uranium, the yield of the power plants, and electricity supply losses) and 1.2 kWh of primary energy to produce 1 final kWh of gas or oil. The primary energy balance sheet is thus heavier for electricity. In terms of primary energy, French consumption was 260 million Teo in 2013. 41% came from nuclear, 30% from oil, 14.5% gas, 10% renewable energy (including hydroelectricity) and 4.5% coal. The energy bill in 2007, before oil prices shot up, came to 45 billion euro, of which 36 billion for oil and 9 billion for gas (sales of electricity brought in 1.8 billion euro). It was 66 billion euro in 2013.

The production of electricity and renewable heat is given in the table below in GWh (mainland France and overseas French territories). The production of these industries depends on climate and therefore varies from one year to the next. For example, the production of hydroelectricity was 64,000 GWh in 2003, with an energy potential of 70,000. Heat pumps draw some energy from the surrounding environment (air, water, land), but energy is lost during electricity production, since power plant yields are below 1. Only systems whose performance coefficient is higher than the inverse yield of the electricity production chain (power plant yields, distribution losses, etc.) can be considered as overall energy producers, but these are rare.

According to a European directive, France should produce 23% of its energy consumption from renewable energy by 2020. On average throughout the world, renewable energy sources represent around 15% of total consumption (20% for electricity).

---

[54]Éric Labouze, *Bâtir avec l'environnement, enjeux écologiques et initiatives industrielles*, Ed. de l'Entrepreneur, Paris, 1993

[55]http://www.statistiques.developpement-durable.gouv.fr/fileadmin/documents/Produits_editor iaux/Publications/References/2014/references-bilan-energie2013-ed-2014-t.pdf

Table 1.3 Production of electricity and renewable heat.

Industry	Electricity	Heat
Hydraulic	59,899	0
Wind	4,106	0
Solar (photovoltaic and thermal)	38	673
Geothermal	95	1,508
Wood energy	1,430	98,368
Biofuel	0	13,607
Total (GWh)	65,568	114,156

1.1.4   Impacts on biodiversity

The number of known species totals around 1.7 million[56], including 1 million animals (750,000 insects) and 250,000 plants. Studies carried out on oceans and tropical forests imply that known species only represent 5% to 35% of the total, which means between 5 and 30 million species.

Some animal or plant species are under threat of extinction, either through direct elimination, or because of human actions that harm their ecosystems. Humans caused 151 higher vertebrate species to disappear over the last 400 years, or one species every 2.7 years. This is 20 times faster than the extinction rate of mammals during past geological eras. According to the International Union for Conservation of Nature (IUCN)[57], out of 41,415 species studied in 2007, 16,306 were under threat of extinction (188 more than in 2006). In total, one mammal in 4, one bird in 8, a third of amphibians and 70% of plants are in peril. France figures among the 10 countries harbouring the highest number of globally threatened species.

The destruction of ombrophilous (rain-loving) tropical forests, which probably contain 75% of total living species and whose service has already been halved[58] (at a rate of 1.8% per year) would step up this rate of species extinction by 1,000 to 10,000 times. For example, the Ivoirian forest has shrunk by 90% in 30 years, and logging is currently carried out throughout the Congo Basin (as far as the borders with Cameroon, Central African Republic, Congo Brazzaville and Gabon). Numerous infractions of logging regulations have been noted: non-respect of cutting thresholds, cutting down too young trees, etc. Deforestation is accompanied by poaching and massive extermination of fauna to supplement foresters' meat. Large mammals (e.g. elephants and gorillas) are chased out of the forests and become a threat to the population, and so killed. French companies, which accomplish 80% of the wood industry turnover in Cameroon, are often put to blame. In Brazil, 80% of Amazonian wood is logged totally illegally, according to NGOs, who use invisible ink that only shows up in ultra-violet light to trace the channels. A certification process, FCE (Forest Stewardship

---

[56] François Ramade, *Écologie appliquée*, Ediscience, Paris, 1995
[57] http://www.iucn.org/
[58] Or 30% depending on the source

Council) was set up in 1993[59]. A forest gains FSC certification if it is managed in line with a certain number of criteria:

- respect for national and local laws, including tax;
- respect for the rights of local inhabitants;
- respect for workers' rights;
- protection of the environment (reduced impacts, preservation of threatened species, erosion control, respect for ecosystems, etc.);
- plantation that takes into account forest regeneration, diversity of species, landscape, etc.
- maintenance of forests and appropriate protection against fire and pests.

The wood from these forests is stamped FSC and carries a code indicating its provenance. Over 100 million hectares are FSC certified in the world, of which 24 million in Canada, 20 million in Russia, 11 million in Sweden, 10 million in the USA, 6 million in Brazil, 5 million in Poland, 3 million in Great Britain and only 18,000 in France. A less-restricting label, PEFC (Programme for the Endorsement of Forest Certification) was set up in 1999 by forest owners in six European countries. The French state has adopted an action plan in favour of tropical forests, requiring 100% certified wood in public orders by 2010.

Wooded areas currently cover 28% of French territory (twice as much as in the 19th century, but half as much as in the Middle Ages), which represents 1.9 billion m^3 of standing timber[60]. The wooded surface area takes up 77% of Finland, 68% of Sweden, 47% of Austria, 35% of Spain, 36% of Portugal, and 30% of Germany and Italy. French annual production is 85 million m^3, but only 35 million m^3 are harvested[61], with a value of over one billion euro (35 billion for all of the wood industry, including carpentry, furniture, paper and cardboard). Collection of wood for heating outside the commercial circuit probably represents 15 to 25 million m^3 per year. About a quarter of the 60 million m^3 consumed is recycled. Domestic wood harvests only cover three quarters of rough wood requirements. Two thirds of the 3.7 million forest owners possess less than one hectare, which tends to increase operating costs compared to north and east European wood.

The building sector consumes high quantities of wood in the form of rough wood (logs, sawn timber), half-finished wood (panels) or finished wood (doors, windows, carpentry parts, parquets, etc.): 3/4 of the volume of resinous timber, 1/7 broadleaved of timber and 2/3 of tropical timber. Half of the European harvest of 200 million m^3 per year is used for firewood. This energy represents close to 10 million tonnes of oil equivalent in France and fuels 7.2 million households.

Impacts on biodiversity also include the destruction of Polynesian reefs because coral sand has been extracted to use as a building material, and the disappearance of Atlantic salmon in numerous rivers (e.g. in Scotland and the Loire) due to dams that prevent them from migrating. Pollution of the oceans with hydrocarbons has caused

---

[59] www.fsc.org
[60] CSTB Magazine, No. 117, September 1998
[61] Isabelle Savini and Bernard Cristofini, Le courrier de l'environnement de l'INRA, No. 42, February 2001

numerous birds to disappear, for example only 100 birds lived in the puffin colony on the Scilly Iles off Cornwall in 1967 compared to 100,000 in 1907. Electric lines are also dangerous for birds: in ten years in France, 700 birds of prey, 220 white stalks and 133 pink flamingos fell victim to electric lines (of which 90% at average voltage and 10% high to very high voltage), according to the French league for the protection of birds[62]. The low average customer density in France (i.e. 50 customers per km² compared to 120 in Germany) makes underground networks difficult to install: only 20% of lines are underground compared to 60% in Germany. Depending on the configuration, the location of installations and the methods used, mortality (expressed as the number of individuals per wind turbine per year) ranges from 0 to 40 for birds and bats, and this aspect is taken into account in impact studies prior to setting up turbines. However, the greatest risks are linked to road traffic, glass-fronted skyscrapers[63] and pesticides (e.g. furathiocarb, imidacloprid, bromadiolone, chloralose, fipronil, etc.).

1.1.5 Proliferation of nuclear power

Nuclear waste treatment is now managed globally (e.g. Japanese waste is processed in The Hague), with all the risks of transporting this kind of material. Unregulated trade by certain countries represents a potential danger. Some sources point to the use of depleted uranium in Iraq to increase projectiles' density, and so their range and efficiency: according to the Institute for Science and International Security in Washington, 940,000 30 mm cartridges (containing 300 g of depleted uranium) were launched from aircraft and 1200 120 mm shells were launched from tanks, i.e. a total of 300 tonnes of depleted uranium[64]. Depleted uranium's radioactivity is half that of natural uranium, but uranium dust produced during explosions can be inhaled, which some doctors consider as toxic.

1.2 REGIONAL SCALE

1.2.1 Effects linked to air pollution

1.2.1.1 Acidification

Sulphur dioxide and nitrogen dioxides generated by human activity are transformed into sulphates and nitrates, sometimes sulphuric and nitric acid in clouds, which then fall down with rainwater. Pollutants directly touch leaves and needles on trees. They alter their protective wax barrier and wash off calcium and magnesium, which is present in the composition of chlorophyll.

However, they can also destroy absorbing roots and wash nutritive elements out of the soil (calcium, magnesium, potassium). The composition of the soil is modified

---

[62] www.lpo.fr
[63] www.flap.org
[64] Le monde diplomatique, April 1995

Sulphur dioxide and nitrogen dioxides generated by human activity are transformed into sulphates and nitrates, sometimes sulphuric and nitric acid in clouds, which then fall down with rainwater. Pollutants directly touch leaves and needles on trees. They alter their protective wax barrier and wash off calcium and magnesium, which is present in the composition of chlorophyll.

Figure 1.3 Tree altered by acidification.

(e.g. increased concentration of aluminium), which, once it crosses a certain threshold, harms some plant species. The cut-off values are given in tables 1.4 and 1.5[65].

In France, despite preponderant west winds, which tend to disperse pollutants, in 1980 the surface area impacted by decline totalled around one million hectares[66] (out of 15 million hectares, representing 27% of the national territory). One quarter of the trees in Europe were damaged (i.e. defoliation above 25%), beech trees and oaks over 60 years old being the hardest hit at close to 60%. The rate of defoliation reached over 70% in the Czech Republic, over 50% in Ukraine, around 40% in Moldavia, Poland, Belorussia and Luxembourg, and over 30% in the Netherlands. Since 1992, the situation has been improving thanks to a European directive regulating the amount of sulphur in fuel and emissions from the most polluting installations (coal-fired electric power plants, large boiler rooms, etc.): for example, French SO_2 emissions went

---

[65] C. Elichegaray, *Retombées des polluants acides: vers des normes draconiennes au 21ème siècle*, Agence de la Qualité de l'Air
[66] François Ramade, *Écologie appliquée*, Ediscience, Paris, 1995

Table 1.4   Critical loads for nitrogen.

Type of vegetation	Critical nitrogen load (kg/ha/yr)
Deciduous trees	5–20
Conifers	3–15
Shrubs	3–5
Meadows	3–10

Table 1.5   Critical loads for total acidity.

Soil class	Examples	Critical acid load (kilomoles $H^+/km^2/yr$)	Sulphur equivalent (kg/ha/yr)
1	granite	<20	<3
2	gneiss	20–50	3–8
3	shale	50–100	8–16
4	gabbros	100–200	16–32
5	limestone	>200	>32

down by 86% (32% for NO_x) from 1980 to 2006[67]. Reinforcing the Gothenburg Protocol, the European Directive of 23 October 2001 on national emission ceilings, incorporated into French law by the Act of 8 July 2003, imposes the following reductions from 2000 to 2010: 43% for SO_2 emissions, 44% for NO_x, 37% for volatile organic compounds (VOC[68]) and 1.4% for ammonia. For France, this means limiting emissions by 1.05 million tonnes of VOC, 315,000 tonnes of SO_2, 810,000 tonnes of NO_x, and 780,000 tonnes of NH_3. Plans to protect the atmosphere have been put in place, and are obligatory in built-up areas of over 250,000 inhabitants in some risk areas.

Boiler rooms with power of over 75 thermies per hour (around 2200 kW) are subject to restrictions on the height of the chimney depending on their power, the rising speed, emission temperature and the amount of sulphur in the combustible[69]. Incinerators and other factories also give off hydrochloric acid (e.g. when incinerating PVC), which adds to the other acids mentioned above.

This impact on vegetation is reinforced by another type of pollution: photo-oxidant pollution through the ozone. The ozone accelerates the oxidation of sulphur dioxide into sulphate, and thus accentuates the acid rain phenomenon. This substance also has

---

[67]CITEPA, Inventory of polluting atmospheric emissions in France in line with the convention on long-distance transboundary atmospheric pollution and the European directive on national emission ceilings, Dec. 2007

[68]VOCs are organic molecules (hydrocarbons, aldehydes, ketones, acids and derivatives), mainly constituted from carbon and hydrogen atoms, but also oxygen, chlorine, sulphur, phosphorus and fluoride. Methane, the main VOC, is non-toxic and practically inert from a photochemical point of view. It is therefore not generally counted, so that we talk of non-methane hydrocarbons (NMHCs): alkanes, isoprene, terpenes, etc.

[69]Serge Lambert et al., *Manuel environnement à l'usage des industriels*, AFNOR, 1994

direct repercussions on human health, some crops, and the ageing of some plastics. It also contributes to the greenhouse effect.

VOC emissions can be reduced by thermal or catalytic incineration, adsorption on active coal or biological processing. The cost of processing depends on the sector of activity, and ranges from 1 to 4 euro per kg of avoided VOC. The reduction of SO_2 and NO_x is linked to energy saving (for heating, transport, etc.) and the choice of fuel: SO_2 is produced during combustion of coal, fuel oil and diesel (which contributes 10%), NO_x come from using fuel in vehicles (75% contribution) and natural gas.

1.2.1.2 Smog and outside air pollution

Polluted atmospheres increase the frequency and gravity of respiratory diseases, among other things.

Figure 1.4 Smog.

The word smog comes from a contraction of smoke and fog, and was coined following a pollution episode in London in December 1952 that caused the death of 4,000 people in 5 days[70]. Coal was still widely used at the time for heating and industry, resulting in high dust and sulphur emission. Winter smog is a phenomenon mainly caused by dust and SO_2, and summer smog is linked to the presence of ozone in the air.

In Europe, 25 million people are exposed to smog in the winter (overshoot of European directives on air quality for SO_2 and matter in suspension) and 40 million to summer smog (overshoot of the World Health Organization's guide values on ozone concentration). According to the Airparif surveillance network, close to 4 million people in the Paris region are exposed all year round to NO_2 concentrations higher than the regulatory 2010 target of $40\,\mu g/m^3$. What is more, 750,000 people are subject to pollution levels caused by benzene that are higher than the French standard of $2\,\mu g/m^3$.

In Paris, the increased presence of NO_2 in the atmosphere is responsible for a 15% to 20% rise in hospitalizations for asthma and home visits for respiratory tract

---

[70]Logan W.P., Mortality in the London Fog Incident, 1952, Lancet, 1953; 1 (7): 336–338

infections[71]. Increased levels of SO_2 have led to a 10% rise in mortality of those suffering from cardiovascular problems. The rise in ozone base levels has caused 20% more hospitalizations of old people suffering from chronic respiratory diseases.

Ozone is produced in the troposphere (low atmosphere, altitude under 10 km) when nitrogen oxides react with VOCs under the effect of ultra-violet rays, and also close to high-voltage lines. The "natural" concentration of ozone is 20 to 50 ppb (parts per billion, i.e. 40 to 100 $\mu g/m^3$), and now sometimes reaches 100 ppb. The thresholds defined in the European directive 2002/3/CE for 2010 are: 110 $\mu g/m^3$ over 8 hours for risks to health (25 days of overrun are currently authorized); 180 $\mu g/m^3$ over 1 hour for informing inhabitants (short-term exposure presents a risk for sensitive people); 240 (average over 1 hour overrun during 3 hours) for alerting inhabitants (short-term exposure presents a risk for the whole population); and 6000 for alteration of vegetation (total from May to July of average hourly levels above 80 $\mu g/m^3$. In France, 65% of VOCs emitted are of human origin ("natural" emissions come from vegetation and forest fires). Non-methane VOC emissions dropped by 51% from 1988 to 2006[72].

Other substances damage air quality: toxic carbon monoxide (CO), which represents between 4% and 6% of exhaust fumes from vehicles[73], NO_x (cf. paragraph 1.2.1.1) and the ozone that these substances indirectly produce (ozone is 500 times more toxic than CO), lead (which mainly comes from road traffic), and especially dust. This involves particles with a diameter of a few microns (pollen particles vary from 10 to 100 μm). Particles with a diameter greater than 10 μm are stopped in the upper respiratory tract[74]. From 3 to 10 μm, they reach the bronchial and bronchiolus tubes. Only particles under 3 μm penetrate into the alveoli of the lungs. Dust particles are generally made up of a carbon nucleus onto which hydrocarbons can be adsorbed. Some compounds can have carcinogenic effects on the lungs, but the main consequence is respiratory disease. The rising death rate is linked to the increase in PM_{10} concentration ($\mu g/m^3$ concentration in particles smaller than 10 μm), and especially $PM_{2.5}$ ($\mu g/m^3$ concentration in particles smaller than 2.5 μm). The concentration of solid matter in the air varies from 0.05 to 0.5 mg/m^3 in rural areas, and 0.1 to 1 mg/m^3 in urban zones[75]. According to a Swedish study[76], the risk of cancer is 50% higher for professionals exposed to diesel motor fumes (machine and motor mechanics, drivers of lorries and building works equipment). The risk is 70% higher for people exposed to soot (chimney sweeps, metal industry, firemen, etc.).

Dust emissions used to be taxed (23 euro per tonne in 2002). ADEME (French Environment and Energy Management Agency) used to collect this tax and use it to support measures for prevention, R&D and measurement (air surveillance network). Emissions were thus reduced by around 29% for PM_{10} and 35% for $PM_{2.5}$ from 1990

---

[71] Study by the Ile de France regional health observatory, published in Energie Plus No. 137, Nov. 94

[72] CITEPA, Inventory of polluting atmospheric emissions in France in line with the convention on long-distance transboundary atmospheric pollution and the European directive on national emission ceilings, Dec. 2007

[73] According to the above reference, CO emissions dropped by 66% from 1980 to 2006

[74] S. Déoux, *L'écologie c'est la santé*, Ed. Frison Roche, 1994

[75] CIBSE, *Guide B: Ventilation and Air Conditioning*, London, 1986

[76] Svensk-Franska forskningsföreningen No. 78, 28 May 1999

to 2006. The tax on air quality, which is now part of the TGAP[77], concerns combustion installations with power higher than 20 MW, installations that emit over 150 tonnes per year of certain pollutants (HCl, hydrocarbons, solvents, VOC, etc.), and household waste incinerators with a capacity of over 3 tonnes per hour. In 2007 it varied from around 40 euro per tonne for VOCs, SO_x and hydrochloric acid, 50 euro for NO_x and 65 euro for N_2O.

France possesses the largest number of household waste incinerators in the European Union, taking all capacities together. Their performance has greatly improved over recent years in terms of limiting emissions (especially dioxins), and energy recovery. Of the 130 installations, over 100 include energy recovery (electricity, heat or co-generation production)[78]. Incineration of chlorinated waste like PVC produces hydrochloric acid and dioxins. Some dioxins are carcinogenic[79]: there are 17 different compounds, of which some are toxic, and the total is generally expressed in TEQ[80]. These substances incur risks of altering foetus growth and diminishing the immune system. Dioxins, which settle around about incinerators, can be concentrated in plants then ingested by animals and concentrated in their flesh. The "disposal" of incinerator cinders along roads contributes to dispersing this pollution. As a result, the rate of dioxins in some milk products exceeds the 1 pg/g limit recommended by the European Union (the alert threshold is 3 pg/g). At this kind of concentration, a child can reach the maximum dose of 1 pg/g per day recommended by the ministry of health.

This is why installations are regulated and dioxin and furan emissions went down by 93% from 1990 to 2006[81], as did PCB (polychlorinated biphenyls: −38%) and PAH (polycyclic aromatic hydrocarbons: −20%). The EU Directive 2000/76/CE of 4 December 2000 limits dioxin emissions from incinerators to $0.1 \, ng/m^3$, a threshold applied in France since early 2006. Thanks to another directive limiting the level of benzene in fuel, the basic concentration measured by Airparif went down by 76% from 1990 to 2005. On the other hand, ozone concentration rose 84% in Ile de France (Paris region) from 1992 to 2005, and overshot the threshold of $110 \, \mu g/m^3$ over 8 hours (which the EU Directive 2002/3/CE says must be respected by 2010) for around thirty days[82].

PVC is used mainly in the building sector (58%, compared to 17% in packaging), yet is only currently collected and recycled as household waste. The PVC recycling rate

[77]Taxe Générale sur les Activités Polluantes (general tax on polluting activities)
[78]Systèmes solaires No. 186, www.energies-renouvelables.org, July–August 2008
[79]The IARC (International Agency for Research on Cancer, which comes under the World Health Organization and the European Union, considers 5 categories: proven as carcinogenic (group 1), probably carcinogenic (group 2A), possibly carcinogenic (group 2B), not classifiable (group 3), and probably not carcinogenic (group 4). Of the 900 substances, mixtures and exposure factors studied, 95 are classed in group 1 (benzine, dioxin 2,3,7,8 TCDD, cadmium, hexavalent chromium, tobacco smoke, asbestos fibres, mineral oil, etc.), 241 in 2B (lead, refractory ceramic fibres, styrene, petrol, etc.), 497 in group 3 and one in group 4
[80]Tetrachlorodibenzo-p-dioxin equivalents, an index obtained by aggregating emissions of different types of dioxins and furans, weighted according to their toxicity
[81]CITEPA, Inventory of air pollutant emissions in France based on the Convention on Long-range Transboundary Air Pollution and the EU Directive on National Emission Ceilings, December 2007, www.citepa.org
[82]www.airparif.asso.fr

is therefore low, and most of the waste is incinerated. As well as chlorine compounds, PVC incineration gives off heavy metals[83] (barium, cadmium) that play a stabilizing and anti-UV protection role. Additives (e.g. phthalate plasticizers, foaming agents, combustion delayers) can represent up to 60% of the weight of PVC compounds.

Coal-fired thermal power plants, incinerators and some industrial procedures generate mercury emissions in the air, which are highly toxic. The widespread use of lead-free petrol, obligatory since January 2000, brought down lead concentration in the air to less than $0.2\,\mu g/m^3$ in 2003 in Paris, twenty times below the 1991 level[84]. Heavy metal emissions dropped overall from 1990 to 2006: ranging from 5% for copper to 97% for lead (77% for cadmium, 71% for mercury, 49% for nickel and 86% for zinc).

The cancer-related death rate has dropped thanks to progress in treatments, but the incidence of the disease (i.e. new cases) rose by 35% in France from 1978 to 2000. 800,000 French people currently live with cancer and 150,000 die from it each year[85]. It is very difficult to identify the proportion of cancers that can be attributed to environmental factors. In France, the Comité de la Prévention et de la Précaution estimates that out of 135,000 cancer cases, 1800 to 5200 could be linked to dioxin[86]. However, the national academy of medicine reports that the incidence of cancer has gone up because of a rising population and more efficient screening[87]. It says that for equivalent population and age groups, the number of cases has gone down. In addition, most cancers are probably the result of tobacco and alcohol consumption (28%), with only 1% of cases linked to pollution. The origin of most cases of cancer remains unknown and the consensus is that research must continue in this area.

In Europe, 15% of couples are sterile, and one child in seven is asthmatic. Allergies are also increasingly common (+15% in 20 years). In the USA, the number of prostate cancers rises by 4% per year, and the risk of breast cancer shot up from 1/20 in 1960 to 1/8 in 2000. Epidemiological studies are still too rare in France.

Reacting to this observation, in 2004, 1000 scientists, of which several Nobel Prize-winners, 1000 NGOs, 200,000 citizens and all the medical association councils of the 25 European Union countries signed the Paris Appeal calling for wider application of precautionary principle for persistent, bioaccumulable or toxic substances; a ban on products whose carcinogenic, mutagen or reprotoxic character is certain or probable; and the substitution of these products, or where impossible, a restriction on their usage[88].

Gaps still remain in our knowledge: of the 100,000 chemical substances on the market[89], very few have been tested for their toxicity and eco-toxicity. However most

---

[83] With density above 5

[84] Idem note 47

[85] Denis Smirou-Navier, *Cancers et environnement: combien en voulez-vous?*, Les annales des Mines, January 2006

[86] Frédéric Denhez, Les pollutions invisibles, Ed. Delachaux et Niestlé, 2005

[87] Académie nationale de médecine et al., Les causes du cancer en France, http://www.academie-medecine.fr, September 2007

[88] French association for therapeutic cancer research, www.artac.info

[89] The European Chemicals Bureau, ECB, listed 100,195 substances in the European Inventory of Existing Commercial Chemical Substances (EINECS) before 1981, and 300 to 350 new substances per year since that date in the European List of New Chemical Substances (ELINCS), cf. http://echa.europa.eu/information-on-chemicals/ec-inventory

of them are commercialized in large quantities: 11% at over 10 tonnes per year (of which 0.6% over 1000 tonnes an 2.7% from 100 to 1000 tonnes), and 61% from 1 to 10 tonnes. Every year, between 300 and 1000 new substances arrive on the market, often without having undergone toxicity tests. The main producing countries are the USA (21%), Germany (19%), Japan (19%), Switzerland (14%), Great Britain (10%) and France (5%).

The Swiss association, Umweltmedizin, which groups doctors concerned by environmental issues, has listed over 65,000 commercial chemical substances (cf. table 1.6). Only a little over 1000 of them have been correctly studied from a toxicity point of view, and the effects of around 45,000 of them are totally unknown. Of those completely or partially studied, 3500 are known to be harmful to health, of which 150 to 200 have carcinogenic effects. A study of the interaction between substances would require considerable resources. As an example, the National Toxicology Program in the USA only has the capacity to test 25 products per year.

Table 1.6 State of knowledge on toxicity of chemical substances.

Type	No.	% full study	% partial study	% little study	% no study
Pesticides	3350	9	27	25	39
Cosmetics	3410	2	24	56	18
Medication	1815	18	21	35	26
Additives	8630	4	16	34	46
Other	48500	0	21	0	79

Faced with this impossibility of ascertaining the toxicity of dangerous substances prior to banning them, the regulation REACH (Registration, Evaluation, and Authorization of Chemicals) required industrials to prove the innocuousness of new products before they could go on the market by December 2010 when substances were commercialized at over 1000 tonnes per year (then 100 tonnes by June 2013 and 10 tonnes by June 2018). Public authorities assess data supplied by industrials when the quantities commercialized exceed 100 tonnes per year. The number of substances concerned is around 10,000 (2300 > 1000 t/yr, 2500 from 10 to 100, 5000 from 1 to 10 t/yr), added to which are 20,000 products commercialized from 1 to 10 tonnes per year, which should be recorded but do not currently require a chemical safety report.

According to the European Commission, the cost of this regulation would be around 3 to 5 billion euro over 11 years, or 0.05% of the turnover of the chemical industry, with benefits in terms of public health totalling 50 billion euro. According to the French chemical industry union, however, REACH would cost 28 billion euro for France alone.

Current regulations only require 30% of substances to be recorded, and only 10% to be studied. Substitution with a less harmful product is not obligatory if "valid control" of the risks is established. Priority is given to substances of "extreme concern", i.e. those that decompose slowly in the environment (very persistent substances); that can accumulate in the body (bio-accumulation); are toxic for humans (carcinogenic, reprotoxic, mutagen, neurotoxic, allergenic, immunotoxic, or capable of altering the hormone system such as endocrine disruptors) or for aquatic environments (substances classed R50/53, with criteria on the biodegradability and lethal dosage for fish, daphnia and algae).

1.2.2 Water abstraction and pollution

1.2.2.1 Water management

Water is mainly found in oceans (97.4%), glaciers (1.9%), underground waters (0.6%), and lakes (0.01%), along with 0.005% in soil moisture, 0.0009% in air humidity, and only 0.00012% in rivers and 0.000008% in living cells[90]. The most striking illustration of non-sustainable water management is probably the Aral Sea, which covered 68,000 km^2 in 1960 but has lost 75% of its surface area and 90% of its volume through use of its two rivers for irrigating cotton fields.

Although hydric resources are not generally under threat in France, water supply can be problematic during certain periods and in some regions. The main usage in France is for cooling electric power plants (21.5 billion m^3, or 64% of abstraction)[91]. Next come consumption for drinking water (17%), irrigation (10%) and industry (10% in quantity, but almost all of the toxic substances emitted and 40% of matter in suspension[92]). Most of the water abstracted returns to the natural environment: net total consumption is 5.75 billion m^3 about half of which is used for irrigation, one quarter for drinking, 23% for electricity production and 4% for industry. Average annual rainfall totals 750 mm, which is a volume of 440 billion m^3, but the stock of instantly available water is on average 100 billion m^3.

In Ile de France, total household consumption represents 2 million m^3 per day, or around 10% of the Seine's average flow. Paris requires a total supply of 3.5 million m^3 per day, against 20,000 m^3 per day in around 1800. In the space of one century, the average daily consumption per person has grown from 7 litres to 165 litres, with strong regional variations (130 litres in Brittany, 240 litres in the Provence Alpes Côte d'Azur region, 150 litres in Ile de France[93]). Tap water consumption is three times higher in Canada. 99% of the French population is currently linked up to the water supply (but 40% of inflows are not protected and 4% of drinking water withdrawals do not meet current standards[94]), while a fifth of the world's population has no access. Four million people die each year from diseases linked to lack of water, and 6,000

---

[90] Jean Matricon, *Vive l'eau*, Ed. Gallimard, 2000
[91] French Ministry for Ecology, Energy, Sustainable Development and Spatial Planning http://www.statistiques.developpement-durable.gouv.fr/fileadmin/documents/Produits_editoriaux/Publications/Chiffres_et_statistiques/2012/Chiffres%20et%20stats%20290%20Pr%C3%A9l%C3%A8vements%20d%27eau%20en%20France%20en%202009%20-%20f%C3%A9vrier%202012.pdf, February 2012
[92] Éric Labouze, *Bâtir avec l'environnement, enjeux écologiques et initiatives industrielles*, Ed. de l'Entrepreneur, Paris, 1993
[93] Institut Français de l'Environnement, 4 pages No. 117, March 2007
[94] Académie de l'eau, symposium organized by former students of Ecole des Mines de Saint Etienne in December 2005, Revue des ingénieurs, March–April 2006

children in the world die each day from having consumed water unfit for drinking, even though such diseases (like diarrhoea) are easy to treat[96].

Water management is carried out per hydraulic basin and town (sometimes by group of basins). European directives regulate the quality of water for different uses (e.g. drinking water, bathing water, piscicultural waters).

1.2.2.2   Dystrophication

Eutrophication, i.e. the enrichment of water by mineral salts[95], develops very slowly in natural conditions (on a geological timescale).

Figure 1.5   Blue-green algae on the coastline, photo Halte aux marées vertes.

However, the significant discharge of fermentable organic matter and phosphate- or nitrate-rich waste by humans considerably speeds up this process, and is termed "dystrophication".

Nitrogen and phosphate fertilizers used in farming, NO_x and dissolved or gassy NH_3 given off by some industrial processes, and phosphate-based washing powders and detergents partially run off towards surface waters (rivers and lakes). They then cause algae to grow and proliferate (e.g. blue-green algae, which covers some beaches) and, in an initial phase, generate oxygen. This effect is exacerbated when trees are cleared from riverbanks along small watercourses, which increases the amount of light and encourages photosynthesis. The abundance of plant biomass reduces the transparency of the water, preventing photosynthetic reactions and the deep release of oxygen. The decomposition of dead algae by micro-organisms ends up starving the milieu of oxygen, thus slowly eliminating all aquatic life. Putrid silt forms on the bed through anaerobic fermentation, giving off hydrogen sulphide and ammonia.

Household phosphate waste can represent over half of the eutrophication indicator of farming waste (for the same population). In France in 1988, 1.4 million tonnes of

---

[95]R. Barbault, *Écologie générale*, Ed. Masson, 1990
[96]www.eaudeparis.fr

detergent were consumed, i.e. 25 kg per person per year, compared to 15 kg in 1975[97] and 29 kg in 2005[98]. Without phosphate removal procedures, water treatment plants only eliminate around 30% of phosphates[99]. Restrictions on tripolyphosphate levels in detergents have therefore been advantageous, since France was the leading consumer of these products in Europe, which were banned from detergents in Switzerland as early as 1986. This restriction has not led to a massive increase in polluting surfactants, since zeolites are harmless clays. The claim that phosphate-free detergent formulas are more eco-toxic than those with phosphate[100] is simply untrue[101]. The additional cost incurred by dystrophication in producing drinking water was estimated to be around 300 million euro in 1999 by the French Water Agencies.

The biodegradability of a detergent, which figures on the packaging, is in fact "primary" biodegradability: i.e. the first stage of decomposition, where molecules are simply broken down into several pieces (e.g. the head and tail of a surfactant). But each of these pieces can still have a harmful effect on the environment. It is only at the final stage of biodegradation that the degradation of totally inoffensive basic molecules (e.g. water, CO_2) takes place. The final biodegradability of plant-based surfactants (such as copra or palm oil) is much better than that of petro-chemical surfactants[102]. Optical brighteners biodegrade very little and some of them are carcinogenic.

Some discharges of pollutants into water directly generate oxygen demand, which can be evaluated by three indicators[103]:

- Chemical oxygen demand (COD): the quantity of oxygen (mg) consumed in a litre of water by oxidizable matter, under the action of a powerful chemical oxidizing agent;
- Biological oxygen demand (BOD): the quantity of oxygen required for oxidation using a biological method (under the action of micro-organisms) of the organic matter present in water; BOD5 is BOD measured over 5 days[104];
- Total oxygen demand (TOD): the quantity of oxygen required for total combustion in an oxidizing atmosphere of one litre of sample.

The combination of dystrophication with other phenomena (in particular a rise in temperature) can result in even lower oxygen levels. Some species (e.g. trout and salmonids) require more oxygen: at under 5 to 6 mg/l they migrate. At under 3 mg/l, fish will die if they cannot migrate. These levels are respectively 3–3.5 and 2 for carp and roach (cyprinids)[105].

---

[97] Que Choisir, *Guide pratique des produits propres*, Nov./Dec. 1990

[98] According to http://www.achatsresponsables-bdd.com/

[99] Agences de l'Eau et ministère de l'environnement, *Lessives, phosphates et eutrophisation des eaux*, 1997

[100] cf. Rhône-Poulenc campaigns in 1989

[101] Roland Carbiener, Report commissioned by the Ministry of the Environment, Strasbourg University, May 1990

[102] Que Choisir, *Guide pratique des produits propres*, Nov./Dec. 1990

[103] AFNOR T90-101, 102 & 103 standards

[104] AFNOR T90-103 standards

[105] Serge Lambert et al., *Manuel environnement à l'usage des industriels*, AFNOR, 1994

Phosphorus also encourages the growth of blue algae or cyanobacteria: microscopic algae that are toxic to humans and animals. Children have been intoxicated and several dozen dogs have died after simply bathing in a river or lake, some bodies of water have thus been closed to bathing when concentration levels have exceeded 100,000 cells per millilitre of water. The ranking of bathing waters is only based on the amount of coliforms (fecal streptococci, escherichia coli): blue for the cleanest, followed by green, yellow and red for the highest levels. From 1976 to 2005, the proportion of bathing locations in conformity rose from 70% to 95%.

The sewage network is linked up to 23.5 million homes, or around 80%[106], but 14,000 towns in France were not connected at the end of 2005[107]. In 2001, the sewage system comprised 250,000 km of wastewater pipes, along with 79,000 km for rainwater evacuation. A little over 5 million households are equipped with a private system. The 1992 water law obliges towns to mark out zones to be connected up to a collective sanitation network, but in 2001, only 42% of towns had done so.

1.2.2.3 Other impacts on aquatic environments

Apart from the problem of dystrophication set out above, pollution can alter transparency (turbidity characterizes the degree of non-transparency, matter in suspension can kill fish by blocking their gills); acidity (increased concentration of CO_2 from bicarbonates in hard water for a pH <6, liberation of molecular ammonia more toxic than the ionic form at pH >8, death of fish if pH <6 or pH >9); temperature (a rise encourages evaporation and fermentation); salinity; and the state of the surface film (drop in gas exchanges).

Sudden temperature changes are dangerous to fish. The maximum tolerable temperature depends on the concentration of oxygen (cf. preceding section): it is 22°C to 24°C for salmonids[108] (the EEC regulatory limit[109] is 21.5°C), and around 30°C for cyprinids (regulatory limit 28°C), if the rise is gradual.

The former Montereau power plant, a thermal 750 MW plant fired by coal (new gas units are under study), used to draw 28 m^3/s from the Seine River, whose minimal flow in that location is barely 30 m^3/s[110]. This means that at low water levels, almost all of the river's water circulated in the condenser. The increase in water temperature was on average 7°C, and could reach 30°C in summertime. Nuclear power plants have a higher power (900, 1300 or 1450 MW for an EPR[111]) and a lower yield (because the temperature of the primary circuit is limited for security reasons), so that the thermal power to be evacuated is much higher on sites that group several sections. Some nuclear stations were thus stopped during the 2003 heat wave when the downstream water temperature became too high (in Switzerland, the temperature of water released must not exceed 30°C).

---

[106]Institut Français de l'Environnement, Les données de l'Environnement, August 2004
[107]Académie de l'eau, symposium organized by former students of the Ecole des Mines de Saint Etienne in December 2005, Revue des ingénieurs, March–April 2006
[108]Serge Lambert et al., *Manuel environnement à l'usage des industriels*, AFNOR, 1994
[109]Directive n° 78/659/CEE du 18 July 1978
[110]François Ramade, *Écologie appliquée*, Ediscience, Paris, 1995
[111]European pressurized reactor

The oxygen content (c, in mg/l) is linked to the water temperature (t, in °C) by the equation:

$$c = 80/(0.2t - 7.1)$$

At around 20°C, the loss of oxygen is therefore 2% per degree of thermal increase. In parallel with this direct impact, the indirect impact (i.e. proliferation of bacteria, increased oxygen demand from all organisms) is also very high. The heat rise also provokes a reduction in calcium concentration. The consequences are reduced diversity of flora (algae and plankton) and fauna (insect larva, molluscs, fish eggs and fries, increase in some fish diseases and mechanical shocks).

The toxicity of some products can be felt in the mid or long term, without necessarily producing immediately visible effects. Indirect contamination (through consuming contaminated species) can lead to a phenomenon of bio-accumulation. Aquatic toxicity is expressed by lethal concentration LC50/24h, i.e. the concentration that causes the death of half of the individuals tested after 24 hours. The opposite of LC50/24h is used as a unit of measurement for inhibiting matter, and expressed in equitox (cf. metox for heavy metals). Human toxicity can depend on age: for example, nitrates are much more harmful to babies than to adults as they transform into nitrites in their stomachs.

Heavy metals

Pollution from heavy metals affects numerous aquatic organisms, through the food chain, animals and humans.

Cadmium is captured by some plants (e.g. lettuce, asparagus, corn and tomatoes). Once absorbed or inhaled, it is stored in the liver and kidneys, where its concentration has increased by 50 times in Europe since the start of the century. It perturbs kidney function and may cause high blood pressure and prostate cancer.

Mercury received media attention at the end of the 1950s, with Minamata Disease, which caused the death of 1022 people[112,113] who had consumed fish contaminated by the Chisso chemical factory (the fish had concentrated the mercury between 100 and 1000 times). The mercury caused a number of impairments and the birth of disabled children. Mineral mercury is not absorbed much, whereas methylmercury, due to its solubility in lipids, penetrates the brain, bone marrow, and peripheral nerves and causes neurological and kidney problems. It was not until 1999 that protection nets enclosing the bay of non-contaminated fish were finally removed after verification that the level of fish contamination in the bay was below official norms.

Another toxic metal (to a lesser extent) is chromium. Cement and preparations that include cement cannot be used or put on the market if, when hydrated, they contain over 0.0002% of soluble hexavalent chromium in the total dry weight of the cement (except in totally automated procedures which include no risk of skin contact)[114]. Lead

---

[112] Actualités Environnement No. 156, 1999

[113] Frédéric Denhez, Les pollutions invisibles, Ed. Delachaux et Niestlé, 2005

[114] Official Journal of the European Union 29-05-2007, annex XVII, restrictions on the manufacture, placing on the market and use of certain dangerous substances, preparations and articles

is even more toxic, for example batteries should be recycled, and regulations must be respected by companies responsible for doing so (companies in the sector were recently condemned by the courts).

One unit of measurement for long-term metal toxicity is metox, in a scale from 1 to 50 according to 4 classes of toxicity[115]: 1 g of zinc or chromium = 1 metox, 1 g of copper or nickel = 5 metox, 1 g of lead or arsenic = 10 metox, 1 g of mercury or cadmium = 50 metox.

How phytosanitary products act on water quality depends on their persistence (i.e. resistance to biodegradation, characterized by their "half life", in other words the time required for half of the product to disappear) and their mobility, evaluated through the soil-water organic carbon partition coefficient (Koc): relationship of concentrations in the immobile (carbon) and mobile (water) phase. The conditions for using the products (number, dates and doses), the composition and texture of the soil, the state of crops and weather conditions (temperature and precipitation) are also important parameters. The maximum content fixed by France for atrazine (used to treat maize) in drinking water is 0.1 mg/l. PCB (polychlorobiphenyl) and DDT (dichloro-diphenyl-trichloroethane) dissolved in water are concentrated in the food chain. As a result, fish in the Baltic Sea are fairly highly contaminated, and a Swedish study carried out on seals[116] established links between ingesting PCB and osteoporosis, an increasingly common disease in Nordic women. Although the USA banned production of PCB in 1976, it persisted in France until 1984 and several rivers are affected: e.g. fishing was forbidden on a 200 km stretch of the Rhône River in 2007.

In order of magnitude, the average daily quantity of pollution per inhabitant is as follows[117]: 90 g of matter in suspension, 57 g of oxidizable matter, 0.2 equitox of inhibiting matter, 1.5 g of reduced nitrogen, 4 g of total phosphorus, 0.05 g of halogenated organic compounds adsorbable on activated carbon, and 0.23 metox.

Water treatment is always carried out in addition to prevention, which is far from the best solution (i.e. cleaner, less wasteful processes, re-use of waste water). It is more expensive: as much as 800 euro per tonne of DBO$_5$ eliminated. The first stages – screening (straining through a simple screen), grit removal, oil removal and sieving – cost from 0 to 2 cents of a euro per m^3.

Neutralization (pH adjustment) lasts around 20 minutes. It is followed by coagulation, which cancels out the electric charges of particles in suspension with mineral reactants (iron trichloride, aluminium sulphate, etc.) in a highly agitated environment (for 20 minutes). Flocculation then causes agglomerated particles to form that are big enough to settle or float. The slow agitation necessary for this process also lasts 20 minutes.

Sedimentation involves slowly passing the waste though a tank (more than one hour) where sludge can settle. Dissolved air is injected at a pressure of 5 bars then quickly brought to atmospheric pressure for around ten minutes, allowing the foam to float and separate. Filtration is used to retain the sludge, which may be thickened by lagooning or on a drying bed: to be discharged they need to be removed by shovel.

---

[115] Serge Lambert et al., *Manuel environnement à l'usage des industriels*, AFNOR, 1994
[116] Association franco-suédoise pour la recherche, Sciences et technologies en Suède No. 116, 28 April 2000
[117] Serge Lambert et al., *Manuel environnement à l'usage des industriels*, AFNOR, 1994

The cost of these operations is around 3 cents per m^3 of water treated (in addition to the previous 2 cents).

Treatment using activated carbon through adsorption deals with diluted solutions (i.e. organic compound concentration lower than 1%) that do not contain salts. The duration of treatment ranges from 20 minutes to 1 hour (optimal) at a cost of 1 cent/m^3. Ultrafiltration consists in filtering particles of several thousandths of a micron through a membrane, without altering the water's mineral quality. It allows separation of matter in suspension (mineral, organic, living – bacteria, virus, fungi, etc.), colloids, emulsions and macromolecules.

Inverse osmosis, used to desalinate seawater, uses membranes with a cut-off of between 0.001 and 0.0001 microns. It modifies the pH and mineral qualities of water, retaining ions as carbonates. Figure 1.6 presents the different filtration methods according to the size of products to be filtered.

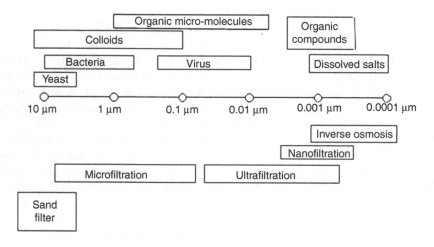

Figure 1.6 Different water filtration methods.[118]

Biological treatments involve feeding micro-organisms with the matter to be eliminated. This can be done in oxidization tanks or using lagooning, but this technique requires large surfaces and a long retention period. It can be done in individual autonomous sanitation systems (e.g. sand filter, spreading trench, macrophyte lagoon: about 30 m^2 for a 6-room house).

In the Ivry treatment plant, water pumped from the Seine River is first filtered on a clay bed that removes the largest particles. Then it crosses a sand bed before passing through filtering tanks where it sits for 6 hours. An ozone treatment removes bacteria and disinfects the water. Final filtration through activated carbon grains completes the process. The whole treatment takes 12 hours, and the production level is 120 million litres of water per day. Water is chlorinated to transport it to four major reservoirs, where it is partially dechlorinated before distribution. Adding chlorine to

---

[118]J. Mallevialle and T. Chambolle, *La qualité de l'eau*, La Recherche No. 221

drinking water considerably reduces the risks of transmitting disease through water (bacteria and viruses). However, chlorine also reacts with organic matter of natural origin present in the water, such as rotting leaves. This chemical reaction produces a family of compounds including trihalomethanes, one of which is chloroform.

1.2.3 Waste and soil pollution

Soil is made up of:

- Mineral matter, which can be classed according to particle size: large elements with a diameter of over 2 mm, sand from 0.005 to 2 mm, silt from 0.002 to 0.005 mm, clay with a diameter under 0.002 mm (2 microns);
- Organic matter (living plants and animals, fresh or decomposing organic matter, humus);
- Groundwater and dissolved elements (organic or mineral);
- Gas constituents (air, gas from organic decomposition and respiration of living beings).

Soil can be used to store water (buffer effect), filter it and purify it (degradation of organic toxic substances through microbial action).

Soil pollution can be diffuse, e.g. radio-isotopes from leaks from nuclear installations, gas encouraging acidification (cf. paragraph 1.2.1), or occasional (industrial or mining installations, landfills, service stations, military installations, etc.).

Heavy metals and other soil pollutants

Pollution from heavy metals, like cadmium, can kill off root systems. Heavy metals tend to be found close to roads and landfills. They are absorbed by plants and dissolved in water, and so enter into the food chain. The main metals involved are lead and cadmium. Lead is mainly released by coal combustion or fuel containing lead, as well as when spreading sludge from treatment plants. Cadmium comes from worn tyres, some heating fuels, and once again fertilization using treatment plant sludge. Its concentration, which is naturally around 2.5 ppm, can be as much as 20 ppm in some sites.

Awareness among the general public is on the rise, but many people still change the oil on their cars and pour the used oil onto the ground, or even abandon old vehicles and domestic appliances in the countryside.

To remove pollution from soil, different treatments can be used depending on the type of polluting substance (halogenated compounds, metals, etc.): biological treatment, immobilization (stabilization using lime for example, or vitrification), isolation-confinement (waterproofing and drainage), dechlorination (by chemical reaction), extraction using solvents (soil leaching), thermal treatment (rotating furnaces) or low-temperature thermal desorption (electrode microwave).

Solid, non-radioactive waste

The concept of waste is purely artificial, since in "nature" products resulting from a process form a raw matter that is potentially useable by another process. This concept, which is actually quite recent, may well disappear in the future.

Several types of waste exist, classed according to their potential impact on the environment and health[120]. Waste is classed as dangerous[121] if it is explosive, inflammable, corrosive, toxic (irritating, harmful, infectious, reprotoxic, carcinogenic, mutagenic, etc.) or if it gives off such substances after transformation, e.g. on contact with water, air, or some other means. The production, storage, transport and elimination of waste are subject to strict regulations. Some products (arsenic, heavy metals) have a very long lifespan.

Some waste is inert (odourless, non-fermentable and non-toxic) and its only downside is taking up space and possibly disfiguring the landscape. One example of inert waste is demolition rubble, as long as it does not contain material like wood or plastic.

Waste that is neither inert nor dangerous is called banal waste. Household waste (and some industrial waste assimilated to this category) can also pollute soil, water and air. The notion of final waste is more recent and refers to waste that, for technical or economic reasons, cannot be either recycled or processed. Since 2002 in France, only this type of waste is supposed to be stored[122]. This date is very late with regards to the 1975 European directive due for application in 1977, and France has been taken to court by the European Commission on this issue.

French waste production was close to 850 million tonnes in 2004[123], of which 374 were agricultural organic waste, 343 from buildings and public works, 90 from companies (6 of them dangerous), 28 household waste (6 of them cumbersome), 14 from local authorities, and 0.2 from healthcare. The building and public works sector produces 295 million tonnes of refuse per year from public works (most of them inert) and 48 million from buildings (cf. Table 1.7).

Table 1.7 Waste from the building industry (millions of tonnes)[119].

Type	Demolition	Renovation	Construction	Total
Inert	29.2	8.6	2.7	40.5
Non-dangerous	1.8	3.5	0.4	5.7
Dangerous	0.2	1.4	0.1	1.7
Total	31.2	13.5	3.2	47.9

Over two-thirds of inert waste is used for embankment construction. It can also be ground up to form reusable granules used under roads for instance. The remaining

---

[119]Institut Français de l'Environnement, 4 pages No. 116, February 2007
[120]Article 5 of Act No. 2002-540 of 18 April 2002 on waste classification
[121]Formerly called "special waste"
[122]Law on waste of 13 July 1992
[123]ADEME, Les déchets en chiffres, 2007 edition

third is put in landfills, mostly former quarries. Recycling is difficult because some materials cannot easily be separated, e.g. plaster stuck to concrete.

Almost half of non-dangerous waste is made up mixed materials, as well as 10% metal, 7% wood, 3% mineral waste (e.g. plaster) and 2% plastic. Significant potential for recycling remains: 95,000 tonnes of metallic waste are not recycled and a considerable amount of wood could be incinerated (30% of wood is still burned on worksites, which is forbidden). 55% of plaster waste (non inert and difficult to recycle) is taken to refuse treatment centres. Dangerous waste is mainly composed of painted or treated wood, asbestos (93,500 tonnes joined to other materials and 8,400 tonnes of sprayed asbestos, which in total makes up almost half of demolition refuse). Asbestos joined to inert materials can be stored in landfills for inert waste or more frequently non-dangerous waste (close to 80%), in specific storage vaults. 5% of asbestos waste has no identified destination and sometimes turns up in rivers.

The following list is an extract regarding construction and demolition waste from Act No. 2002-540 of 18 April 2002 on waste classification (following the EU classification).

17 00 00 Construction and demolition waste (including excavated soil from contaminated sites).
  17 01 00 Concrete, bricks, tiles and ceramics.
  17 01 01 concrete.
  17 01 02 bricks.
  17 01 03 Tiles and ceramics.
  17 01 06 mixtures of, or separate fractions of concrete, bricks, tiles and ceramics containing dangerous substances.
  17 01 07 mixture of concrete, bricks, tiles and ceramics other than those mentioned in 17 01 06.
17 02 00 Wood, glass and plastic.
  17 02 01 wood.
  17 02 02 glass.
  17 02 03 plastic.
  17 02 04 glass, plastic and wood containing or contaminated with dangerous substances.
17 03 00 Bituminous mixtures, coal tar and tarred products.
  17 03 01 bituminous mixtures containing coal tar.
  17 03 02 bituminous mixtures containing other than those mentioned in 17 03 01.
  17 03 03 coal tar and tarred products.
17 04 00 Metals (including their alloys).
  17 04 01 copper, bronze, brass.
  17 04 02 aluminium.
  17 04 03 lead.
  17 04 04 zinc.
  17 04 05 iron and steel.
  17 04 06 tin.
  17 04 07 mixed metals.
  17 04 09 metal waste contaminated with dangerous substances.
  17 04 10 cables containing oil, coal tar and other dangerous substances.

17 04 11 cables other than those mentioned in 17 04 10.

17 05 00 Soil (including excavated soil from contaminated sites), stones and dredging spoil.

17 05 03 soil and stones containing dangerous substances.

17 05 04 soil and stones other than those mentioned in 17 05 03.

17 05 05 dredging spoil containing dangerous substances.

17 05 06 dredging spoil other than those mentioned 17 05 05.

17 05 07 track ballast containing dangerous substances.

17 05 08 track ballast other than those mentioned in 17 05 07.

17 06 00 Insulation materials and asbestos-containing construction materials.

17 06 01 insulation materials containing asbestos.

17 06 03 other insulation materials consisting of or containing dangerous substances.

17 06 04 insulation materials other than those mentioned in 17 06 01 and 17 06 03.

17 06 05 construction materials containing asbestos.

17 08 00 Gypsum-based construction material.

17 08 01 gypsum-based construction materials contaminated with dangerous substances.

17 08 02 gypsum-based construction materials other than those mentioned in 17 08 01.

17 09 00 Other construction and demolition waste.

17 09 01 construction and demolition wastes containing mercury.

17 09 02 construction and demolition wastes containing pcb (for example pcb-containing sealants, pcb-containing resin-based floorings, pcb-containing sealed glazing units, pcb-containing capacitors).

17 09 03 other construction and demolition wastes (including mixed wastes) containing dangerous substances.

17 09 04 mixed construction and demolition wastes other than those mentioned in 17 09 01, 17 09 02 and 17 09 03.

In 2004, household waste production was 353 kg per inhabitant per year (twice as much as in 1960, but slightly less than in 2002), of which 290 kg were buried or incinerated. The French ministry of ecology and sustainable development aims to reduce this quantity to 250 kg per inhabitant per year by 2010 and 200 kg by 2015[124]. According to ADEME statistics[125], 23% of household waste is packaging. This is made up of 29% organic matter, 25% paper/cardboard, 13% glass, 11% plastic, 6% textiles, 4% metal, and various other materials of which 0.5% is special waste. The European Directive on Landfill of Waste imposes giving priority to alternatives like recycling and compost. Targets include reducing biodegradable municipal waste in landfills by 75% by 2006, 50% by 2009 and 35% by 2016 compared to 1995 quantities.

Quantities of household waste produced in 2005 and the various treatment methods are presented in the table below. The percentage of waste incinerated has gone up

---

[124]ADEME, *Prévention de la production des déchets*, 2006 review

[125]ADEME, *Déchets municipaux: les chiffres clés*, February 1998

Table 1.8 Household waste treatment methods in france (2011).[126]

Treatment	No. of installations	Quantities received (millions of tonnes per yr)	%
Landfill	244	9.7	28
Incineration with recovery	114	12	34
Incineration without recovery	15	0.4	1
Compost	600	5.7	16
Sorting	366	7.2	21
Total	35	100	

slightly over recent years, but the number of incinerators has halved: small installations have been closed, and the focus is on energy recovery. When a factory is not too far from an urban zone, energy recovery can be channelled into a district heating system feeding into buildings, and even produce electricity as well (cogeneration). In rural areas, electricity production is the only type of energy recuperation that is worthwhile for the incineration industry. In France, 99% of the population benefits from household waste collection and almost 95% from selective collection of several materials. Almost 3,500 refuse treatment centres also exist.

Prevention strategies (clean technology, waste recovery) should be implemented as a priority. In addition, the main refuse treatment channels are:

- Thermal means (incineration, preferably with energy recovery, pyrolysis[127] or plasma); the latter technique makes it possible to reach a maximum destruction rate: all matter is burned, and the mineral residue (heavy metals, etc.) is vitrified, forming inert waste (cost is however high: around 600 euro per tonne);
- Physico-chemical means (pH adjustment, precipitation or oxygen-reducing neutralization of the aqueous phase, depletion by gravity, or mechanical dehydration by filtering sludge waste);
- Biological means (compost, fermentation, methanization);
- Solidification (vitrification, coating in a mix of polymer, silicate, cement, extraction using solvents of organic micropollutants, neutralization, precipitation or polymerization, possibly by coating with other waste).

Industrials are increasingly efficient at treating dangerous waste. Some waste is even recycled during processing, for example to use as a combustible in cement factories. The temperature is 850°C in a standard incinerator, 1100°C in an incinerator used for chlorine concentration of above 1%, and 2000°C in a cement furnace, which can process e.g. animal meal, tyres and used oil.

---

[126]http://www.statistiques.developpement-durable.gouv.fr/lessentiel/ar/286/1154/traitement-dechets.html
[127]Decomposition of a body under the effect of heat without any other element coming into play (e.g. oxygen for combustion)

Overall, energy recovery from waste produces 3650 GWh of electricity and 7580 GWh of heat in 2013[128], but also emissions into the air and over 2 million tonnes of clinker[129]. The composition of clinker is regularly controlled (leaching test and measure of unburned refuse), in order to ensure that the maximum levels of heavy metals (cadmium, chrome VI, lead, arsenic and mercury), sulphates, organic compounds and unburned refuse are respected. Depending on these levels, clinker can be directly recuperated (category "V"), or "matured" (category "M": storage for at least 2 months and at most a year, during which time exothermal chemical reactions take place) or even stored permanently in landfill because of their high leachable fraction (category "S").

Possible uses of clinker include road and parking structures (except for reservoir or porous pavements) or compacted embankments (except trench embankments that include metal pipes or draining systems). Worksites must be located over 30 m from a watercourse, outside flood zones or protection perimeters close to drinking water catchment areas, and at a sufficient distance from the highest known water level.

Some types of dangerous waste can be received in class 1 sites, i.e. impermeable sites. Waste is not accepted when it is liquid or contains biocides, explosives, radioactive material, or soluble heavy metal salts. Banal waste, which can be considered as household waste, can be put in class 2 landfills (semi-permeable sites). This excludes waste that may percolate pollutants or soluble salts, even if not toxic, into water (i.e. water that passes vertically through the landfill). This information is attained through analyzing the leachate[130]. The classification of waste requires measuring numerous parameters (pH, conductivity, COD, content of heavy metals, cyanide, phenols, sulphates, anionic detergents, arsenic, organochlorine pesticides, PCB and phtalates, hydrocarbons). Class 3 landfills (permeable sites) receive inert waste.

Water that passes through waste catalyzes chemical reactions Household waste is still sometimes put in permeable landfill sites, but this practice is now banned in France. When they are located close to the water table, landfills can have a harmful effect on water quality. (aerobic and anaerobic fermentation with methane production) and conveys the dissolved substances. It thus creates a major risk of pollution, either in the water table or in springs and streams located on the site or close by.[131] Drainage is then required: upstream runoff must be collected to reduce the polluted flow, and the leachate (polluted water that has passed through waste) must be collected downstream to avoid infiltration into the soil. First, the subsoil must be verified to make sure it is impermeable to avoid polluting the water table. Then existing streams (both permanent and intermittent) need to be channelled to protect clear water. Superficial drainage is

[128] http://www.enr.fr/valorisation-energetique-des-dechets
[129] Jean-Michel Laudrin, *Les mâchefers d'incinération d'ordures ménagères*, Génie Urbain, November 1994
[130] AFNOR standard X31-210, leaching is a phenomenon involving water-soluble elements or other solvents
[131] Georges Gimbert and Christiane Dufour, *Le traitement des lixiviats*, Génie Urbain, November 1994

Household waste is still sometimes put in permeable landfill sites, but this practice is now banned in France.

When they are located close to the water table, landfills can have a harmful effect on water quality.

Photo UNEP.[132]

used to lead the runoff towards a clean water system. Lastly, internal drainage captures the polluted waters and takes them to a water treatment plant. Treatment involves several stages: de-silting, fractioned distillation at 65°C, fermentation for 72 hours, and sedimentation. The methane produced must be recuperated, and possibly used in the actual purification (distillation).

A European directive, which should have been applied in 2001, states that the base of landfills must be waterproofed, leachate must be collected and gas emissions must be checked. Four main phases are set out: a short, initial aerobic phase, an anaerobic acidogenic phase (of variable length, longer than the 1st phase), an anaerobic methanogenic phase (up to several centuries), and a final aerobic phase.

IFREMER (Institut Français de Recherche pour l'exploitation de la Mer) carried out a study on waste abandoned on the coast, whose quantity rose from 1.7 kg to 6.5 kg per linear metre from 1982 to 1995. 70% of this waste comes from watercourses, 20% from boats and 10% from holidaymakers. The study also provides an estimation of the lifespan of the waste (see table below).

Reclaiming matter contained in waste is a good thing as long as the matter is sound and decontaminated[133]. It should not be a pretext for getting rid of pollutants by simply diluting them in final products and consumer staples. Controversy has arisen following the introduction of certain heavy metals (batteries, etc.) into steelworks converters, organochlorines into cement, radioactive waste into glass wool, and arsenic into lime, etc. Currently, 80% of non-dangerous waste is reclaimed from companies with over ten employees.

Compost is sometimes used in rural areas, especially for plant waste (mowing, pruning, etc.): 1.73 million tonnes of compost are produced each year[134]. A mix of household and plant waste facilitates composting. When it comes to recycling,

---

[132]United Nations Environment Programme

[133]Marc Lassus, *La fausse valorisation ou les excès du recyclage*, Environnement et Technique, No. 207, June 2001

[134]ADEME, Les déchets en chiffres, www.ademe.fr, 2006

Table 1.9 Persistence of different types of waste.

Type of waste	Estimated lifespan
Apple core	Several days to 6 months
Paper tissue	3 months
Bus ticket	3 to 4 months
Hemp rope	3 to 4 months
Cigarette end without filter	3 to 4 months
Cigarette end with filter	1 to 2 years
Match	6 months
Chewing gum	5 years
Plank of wood	13 to 15 years
Iron food tin	10 to 100 years
Disposable lighter	100 years
Aluminium can	200 to 500 years
Plastic bag	450 years
Plastic bottle	100 to 1000 years
Glass bottle	4000 years

5.6 million tonnes of packaging were recycled in 2005[135], representing 53% of the total (64% counting energy recovery). 57% of metals are recycled, 47% of paper-cardboard[136] (compared to 74% in Germany), 19% of plastic packaging, 60% of glass containers[137] and 21% of wood.

In France, paper is made from 60% used paper and 40% wood pulp. Used paper is therefore not only employed to make recycled paper. Thus, even to produce "standard" paper, we import used paper from Germany, Belgium, the Netherlands and the United Kingdom (close to 500,000 tonnes in 2000[138]). Recycling used paper requires less energy than making wood pulp and avoids impacts brought about by its production (i.e. water consumption, organic and chemical waste). One tonne of used paper is sufficient to produce one tonne of pulp, thus avoiding consuming 2 to 3 tonnes of wood. However, other types of pollution are involved: 20% of recycled paper is deinked, which releases heavy metals that end up in treatment plants. Whitening using chlorine gives off organochlorines: 4 kg per tonne of pulp, including dioxins.

Magnets in treatment centres (around one third of centres were equipped in 1990) are used to sort scrap iron that, once ground, cleaned and purified, can present an iron content of 92%. Steel can also be obtained by removing metal from incineration cinders, but with losses due to oxidization during combustion. Producing steel from this recuperated scrap metal requires half as much energy as manufacturing from the mineral (2500 thermies per tonne produced[139]). Aluminium can also be

---

[135] ADEME, La valorisation des emballages en France, June 2004
[136] www.cercle-recyclage.asso.fr
[137] www.verre-avenir.fr, 2007 data
[138] www.cercle-recyclage.asso.fr
[139] Que Choisir, *Guide pratique des produits propres*, Nov./Dec. 1990

recuperated using the eddy current technique. Recycling provides a large proportion of the steel consumed in France, 68% of lead, about 25% of copper and zinc and 30% of aluminium.

From 277,000 tonnes in 1979, the quantity of recycled glass rose to 2 million tonnes in 2006[140]. Each tonne of used, ground and melted glass saves 100 kg of fuel, 700 kg of sand, 280 kg of limestone and 230 kg of soda ash[141] compared to producing it from scratch. Cullet (collected, broken glass) is used in its manufacture in a proportion ranging from 30% to 60% (36% on average), and allows a reduction in the fusion temperature.

Around one million tonnes of PVC are thrown into household and banal waste each year in Europe, and the same quantity is present in building and demolition waste. Most of this waste is sent to landfills, but 600,000 tonnes are incinerated. Less than 3% are recycled, which is made difficult by the presence of various heavy metals used as stabilizers, but the PVC industry is committed to increasing this proportion. Chemical recycling (fragmentation of polymer molecules) or physical recycling (dissolution-precipitation) is sometimes proposed as an alternative to mechanical recycling (shredding, screening and grinding).

The use of PVC as a combustible in blast furnaces or cement plant ovens can impact negatively on the quality of iron or cement produced and corrode equipment due to the formation of HCl. A reduction of the use of combustible by several percent is under study. PVC is responsible for around half of the chlorine generated by household waste incinerators. Hydrochloric acid must be neutralized to respect thresholds linked to reducing acid rain. The incineration of one kg of hard PVC generates between 0.8 kg and 1.4 kg of residue from smoke treatment depending on the procedure, sometimes in addition to 0.4 kg to 0.9 kg of liquid effluent. Flexible PVC generates about half that amount, depending on the percentage of plasticizer. The residues contain heavy metals and dioxins and are thus considered as dangerous waste and put into landfills, where they produce leachate. The additional cost of incinerating PVC is estimated at between 20 to 300 euro per tonne depending on the procedure. This cost is not borne by the PVC industry, but comes into the overall cost of incineration.

PVC in landfills releases phtalates and heavy metals into the leachate, as well as the products from their degradation. The loss of phtalates can also contribute to gas emission from landfills.

Around 30% of old clothes collected are exported and reused; 30% are employed in industrial cleaning up; almost 20% are unthreaded and used again to make new textiles; 10% are used in paperboard mills, more specifically to make bituminous cardboard, thermal and sound insulation; the remaining 10% are eliminated[142]. Each year, 330,000 tonnes of textile are collected, and each inhabitant buys 15 kg of new

---

[140]Institut d'informations et de conjonctures professionnelles (statistics observatory of FEDEREC, the French federation of recycling companies), April 2007
[141]Chambre syndicale des verreries mécaniques (mechanical glassworks union chamber), 1990 figures. A more recent source indicates savings of 85 l of fuel and a reduction of 240 kg in CO_2 emissions, soda ash no longer being necessary
[142]Que Choisir, *Guide pratique des produits propres*, Nov./Dec. 1990

Table 1.10 Volumes of waste collected by recycling companies (2006).[143]

Type of waste	Volume collected in 2006 (Mt)	Progress from 1999–2006 (%)
Scrap iron and OUV*	14.6	+34
Non-ferrous metals	1.7	+31
Paper-cardboard	5.6	+23
Textiles	0.33	+30
Plastic	0.36	+62
Glass	2.11	+32
Mixed BIW**	7.9	+32
Palettes	66.6***	+39
Other	7	+39
Total (except palettes)	39.6	

*Out of Use Vehicles; **Banal Industrial Waste; ***Millions of palettes

clothes (900,000 tonnes in total). The Table 1.10 shows the volumes of waste collected in 2006 and progress compared to 1999.

France consumes around a billion batteries a year (quantity declared on the market), or about 30 tonnes[144] including 7 tonnes of button cells[145]. Some alkaline batteries have a low mercury content (i.e. under 0.02%), others contain no cadmium. Some saline batteries are mercury-free. "Button" cell batteries can contain a high percentage of mercury (up to 30%). France was very late in applying the 1991 European directive obliging member states to collect and recycle batteries containing metals harmful to the environment, and in 1996 it was ordered to pay ECU 150,000 per day of delay. Only 30% of batteries are currently collected, even though they simply need to be taken to a local store. Car batteries can be taken to a refuse centre, where the lead can be reclaimed.

The cost of land disposal for banal waste varies from EUR 8 per tonne in Spain to 200 in Germany. It is on average EUR 54 before tax in France, but with strong variations (from EUR 11 to EUR 111 per tonne)[146]. The cost of incineration is about the same: from EUR 50 to 120, but selective waste collection is more costly than regular collection: EUR 140 per tonne compared to EUR 70). Sometimes, the cost of taking waste to a grouping centre must be added (around EUR 20 per tonne). Dangerous waste treatment is much more expensive: around EUR 250 for high-temperature incineration[147] and up to EUR 1000 per tonne[148] for other procedures. Treatment

---

[143]Institut d'informations et de conjonctures professionnelles (statistics observatory of FEDEREC, the French federation of recycling companies), April 2007

[144]ADEME et Vous No. 11, December 2007–January 2008

[145]Que Choisir, *Guide pratique des produits propres*, Nov./Dec. 1990

[146]ADEME, Le prix de la mise en décharge des déchets non dangereux gérés par les collectivités, March 2006

[147]ADEME, Les marchés des activités liées aux déchets, situation 2005/2006 et perspectives 2007

[148]Rudologia, Gestion des déchets dangereux en collectivités, Forum déchets, Association des Maires de France, November 2004

of inert waste is a lot cheaper if it is sorted (around EUR 10 per tonne) rather than mixed (about EUR 100 per tonne). In 2004, France spent 11 billion euro on waste treatment, including 1.6 billion on investment (particularly upgrading incinerators) and 9.5 billion on overheads, of which 59% was funded by household bills and local authorities.

1.2.4 Radioactivity

The sievert and the rem are units of measurement used to evaluate the biological impact of radiation over a specific time period[149]. Physical doses are expressed in gray (gy) or rad (1 rad = 0.01 gy). One gray corresponds to 1 joule per kg of matter. The equivalent biological dose is calculated according to the physical dose absorbed, taking into account the nature of the radiation, its energy and distribution over time and space. For X and γ (photons) rays and β radiation (electrons), 1 Sv is equal to 1 gy, and 1 rem to 1 rad. For α rays, 1 rad (physical dose) is equal to 10 rems (biological effect). For neutrons, 1 rad equals between 3 and 10 rems depending on their energy. Dose flow is expressed in gy/h for a physical dose or Sv/h for a biological dose.

Activity represents the number of disintegrations of radioactive bodies per second. The Becquerel (Bq) equals one disintegrated atom per second, one Curie (Ci) is worth $3.7\ 10^{10}$ Bq. The Euratom 96/29 directive considers $1.11\ 10^{-8}$ Sv as equivalent to one Bq. The effect of radioactive particles is very different in the case of external contamination (exposure to radiation) and internal contamination (inhalation or ingestion of matter). When plutonium 239 dust with a diameter of just 1 μm reaches the lungs it generates highly ionizing temporary irradiation, which usually leads to cancer even at very low doses.

No threshold exists below which radiation is inoffensive[150]. The impacts of low doses are however very difficult to ascertain since they become apparent after a latent period that may be several years (leukaemia), tens of years (solid tumours) or several generations (genetic disease).

Natural radiation at sea level in Europe is around 0.03 μSv/h, and increases with altitude and latitude. The presence of radon released from some land (especially granite-based) can increase the dosage received by people living in a building (cf. paragraph 1.4). Nuclear arms testing in the atmosphere, leaks and accidents at nuclear installations have increased the level of radioactivity, which is around 0.10 μSv/h in several European towns[151]. Natural radioactivity explains why some ashes released from wood combustion have slightly exceeded the base level, in particular potassium 40 concentration, but without any risks to health according to the nuclear safety authorities.

---

[149]Groupement de scientifiques pour l'information sur l'énergie nucléaire (Association of Scientists for Information on Nuclear Energy, GSIEN)

[150]S. Muratet and O. Vernet, *Etude et comparaison des effets génétiques et biologiques de la pollution atmosphérique et de très faibles doses de radioactivité naturelle*, Report by P. Sabatier University, Toulouse, 1989

[151]Seminar on urban energy and environmental indicators, Rennes, February 1994

The impacts of the nuclear industry can be local (proximity to uranium mines, stores of depleted uranium, installations for manufacturing fuel and reprocessing, power plants and waste storage sites) or regional (e.g. Chernobyl accident), and the risks can be global (transport or use of plutonium, risk of proliferation). According to the International Atomic Energy Agency (IAEA), the Chernobyl accident caused 36 deaths, and 4000 people could die following exposure to radiation after the accident. A brochure produced by the United Nations Office for the Coordination of Humanitarian Affairs reports that 31 people died in the first three months following the accident, plus 106 closely involved workers since that date, and 600,000 operators were highly contaminated (e.g. soldiers, government agents and other civil operators sent in just after the explosion to neutralize the reactor and bury contaminated waste). According to some sources, 25,000 of these "liquidators" died following the catastrophe, and 200,000 became disabled[152]. 75 million people living in Ukraine, Belarus and Russia were exposed to different degrees (135,000 within a 30 km radius people were evacuated, 350,000 people in total were displaced) and the numbers of deaths from radiation-induced cancer could represent tens of thousands of cases in 70 years.

In France, cancer registers are not systematically kept: it was not until 14 years after the "cloud" passed that the Rhône Alpes region set up a register for thyroid cancer. Mushrooms concentrate numerous toxic substances (e.g. heavy metals), and in particular radioactivity (caesium 137, iodine with a low lifespan). The underground part of mushrooms, mycelium, is formed from very thin filaments located several centimetres deep. These filaments form a wide network (from several dozen to several hundred m^2) and can prosper for decades, concentrating up to 30% of the caesium in the ground, whereas it only represents 5% to 6% in mass. The most contaminated species (pied de mouton, chanterelles, bolets) have exceeded the limit of 600 bq/kg set by the European Commission for commercializing products.

The risk of nuclear accidents is obviously much lower with the technology used in western countries, but it exists even so. It is thus difficult to insure nuclear power plants against the risk of major accidents and a law has to be voted for the state to take responsibility for the cost of damages.

Radioactive waste with low and average activity (class A) is scheduled to be stored for 300 years in above-ground storage centres (in La Hague in the Manche area, and Soulaines-Dhuys in the Aube). After these 300 years, the sites will be returned to normal, despite long-term residual emissions (i.e. residual radioactivity after ten periods, tolerance of the existence of alpha emitters in class A waste).

To date, no satisfactory solution has been found for the long-term storage of nuclear waste. In France, production is around 0.5 kg per person per year[153]. Experiments *in situ* are underway or planned to study underground storage in different types of soil. Some recommend surface storage since it is easier to monitor. The risk of underground storage is geological, since an unanticipated event could disturb

[152]Groupement de scientifiques pour l'information sur l'énergie nucléaire, Gazette nucléaire No. 225–226, November 2005
[153]ANDRA, *La gestion des déchets radioactifs*, 1996

stocks[154]. Above-ground storage presents risks relating to society, durability and stability for surveillance and covering costs; the risks of conflict or attacks should also be considered.

The cost of long-term storage depends heavily on the capitalization rate considered. At an 8% rate, the cost of surveillance becomes negligible after 200 years. However, it is difficult to assess economic growth over a long period. The rate was only around 2% over the last 200 years. Around twenty periods need to pass before waste is sufficiently deactivated to become danger-free, i.e. 490,000 years for plutonium 239.

In September 1996, the Office de Protection Contre les Rayonnements Ionisants (OPRI), a state body under the Ministry of Health, authorized ISOVER Saint Gobain to use radioactive sodium silicate in the manufacture of glass wool. Production commenced during the summer of 1997 at the company's factory in Orange, but despite the product's low toxicity, it was obliged to resume standard production due to customer concerns.

I.2.5 Risks

Some operations – worksites, diverse processes (chemistry and petrochemistry, biotechnology, nuclear) – and the transportation of dangerous matter can carry risks that, although improbable, could have a non-negligible overall impact. Classified facilities are subject to numerous regulations, like the European "Seveso Directive" (Council Directive No. 82/501/EEC of 24 June 1982), transcribed into French law in July 1985, which led to the classification of over 320 installations. These can be subject either to authorization, or simple declaration.

Numerous regulations cover all aspects of environmental law: democratization of public enquiries, protection of the environment, urban code, civil security (operation plans inside the factory, specific intervention plans – level 1 non-toxic, level 2 internal, level 3 external – the Orsec plan at borough, regional or national level).

Danger studies and impact studies are obligatory for classified facilities (and sometimes safety studies, which are even more demanding and checked by a third party, for the most dangerous installations). Brochures inform the public in zones coming under emergency plans, with radio messages in case of alert, and periodical publications of reports describing the state of the environment[155].

Even when they pose no risk to the neighbouring population, some industrial procedures jeopardize the health of staff, especially in countries where occupational health regulations are less strict (e.g. the use of asbestos in Brazil, and not so long ago in France). The building site can present a risk in itself, especially when it comes to flooding, landslides and forest fires.

Transportation of dangerous matter is also regulated by a set of national and international laws for each mode of transport (rail, road, sea and air). Road transport

---

[154]Ghislain de Marsily, quoted in Les cahiers de Global Chance No. 22, November 2006
[155]European directive No. 90/313/EEC of 7 June 1990: public authorities must make available information they hold regarding the environment to anyone who requests it without justification; member states can refuse to give out certain information, but they must justify their refusal; the authorities' response must be given within two months, and the decision can be the object of a litigation claim

presents non-negligible risks: lorries make up 2% of vehicles but are responsible for 13% of deaths on the road[156].

Table 1.11 Occupational accidents recorded in 2006.

Activity	No. of employees	Permanent incapacities	Deaths
Engineering + architecture + hygiene/safety + surveyors	140136	92	2
Construction private houses (PH)	35686	259	2
General companies except PH	192455	1732	21
Earthworks and foundations	63544	375	17
Masonry and heavy work except PH + stone cutting + marble work	117812	886	8
Carpentry + glass + blinds + shutters	114506	944	13
Roofing + roof frame + proofing	72317	686	16
Plumbing + heating + chimneys + sweeping	145085	776	9
Plastering + insulation + asbestos remov.	57322	407	2
Electricity + aerials	136206	592	9
Painting + surfacing + fitting + decoration	174449	1017	18
Metalwork + locksmith	46172	424	5
Scaffolding, maintenance and hiring of equipment	17743	127	6
Demolition	5029	56	3
Total building	1318462	8373	131
Public works	168805	1123	26
Total	1487267	9496	157

Transporting hydrocarbons brings the risk of oil slicks: 525 accidents from 1975 to 1980, i.e. an average 440,000 tonnes per year. Accidents in offshore wells and fuel expelled from ships add to these figures, with the total volume of oil released representing around 3.6 million tonnes per year. As one tonne spreads over $12\,km^2$, the ocean is covered in a thin film of hydrocarbon[157]. Oil slicks cause the death of between 150,000 and 450,000 birds a year in the North Sea and the North Atlantic. To give an idea of size, the Erika, which was transporting 20,000 tonnes of hydrocarbon at the time of its accident in 2000, lost around half of its cargo, killing 12,000 birds (the Amoco Cadiz released 230,000 tonnes in 1978). The cost of the accident for the taxpayer is an estimated 1.2 billion euro).

Oil is made up of a great number of different compounds (as many as 3000), of which only 250 have an identified molecule. 95 of them have been the object of a toxicity study and only 25 have been thoroughly studied. The lightest molecules dissolve more easily, so that fuel reaching a beach is mainly made up of heavy molecules, which are generally more carcinogenic (e.g. benzoapyrene). Pipelines are also the source of pollution, e.g. in Siberia, 15,000 tonnes of fuel flow into the sea each year.

---

[156] French Ministry of Ecology, Energy, Sustainable Development and Spatial Planning, La revue Sécurité routière No. 133, www2.securiteroutiere.gouv.fr, June 2003
[157] François Ramade, *Écologie appliquée*, Ediscience, Paris, 1995

The building sector is one of the most exposed to risk: with 7% of the country's employees (1.3 million), it notches up 23% of the deaths, 21% of accidents with permanent incapacity and nearly 20% of lost work days. The country's national health insurance fund for salaried workers reported 100,000 accidents with work stoppage in 2006, of which over 6000 were cases of permanent invalidity and 131 were deaths[158] (267 in 2000). The table below summarizes the data by type of job. The highest accident rates are observed in demolition and roofing.

The act of 5 November 2001 requires employers to establish a "single risk evaluation document" setting out prevention measures, but some companies are slow to take up the measure.

1.3 LOCAL SCALE

1.3.1 Noise

Sound is the result of a variation in atmospheric pressure, caused by a sound source. This variation is called acoustic pressure. Sound levels are expressed in decibels (dB) on a logarithmic scale and defined in relation to a reference: the threshold of auditory perception for a frequency of 1000 Hertz. Human perception depends on the frequency. Weighting is applied to obtain a filtered sound level in dB (A).

The level of noise pollution is regulated: the values depend on the site (i.e. hospital zone, rural residential zone, urban or suburban zone, commercial or industrial zone) and the time of day. Regulations also take into account the presence of pulsed noise and sounds with a marked tonality ("pure" frequency). Emergence is the difference between the reception sound level and the initial level (measured when the installation is at rest).

Noise apparently participates in 11% of occupational accidents and 15% of lost working days[159]. It also has an incidence on the quality of work, and on the quality of life in the home. Deafness is the leading occupational disease, representing 50% of the total compensation budget[160]. Thirteen million French people live in zones where the average general noise is between 55 and 65 dB (A), six million people are exposed to daytime outside noise levels between 65 and 70 dB (A), which is the physiological reaction threshold (modification of cardiac and respiratory rhythms, skin temperature, digestive system, brain activity, etc.). Auditory fatigue is a reversible phenomenon, unlike deafness. The latter can be caused by prolonged exposure to high sound levels (90 dB (A)), or short exposure to very high levels (140 dB (A)). Table 1.12 gives an overview of the different sound levels[161].

---

[158]Caisse Nationale de l'assurance maladie des travailleurs salariés, technological statistics of occupational accidents, results for the national technical building and public works committee, 2006

[159]Serge Lambert et al., *Manuel environnement à l'usage des industriels*, AFNOR, 1994

[160]Éric Labouze, *Bâtir avec l'environnement, enjeux écologiques et initiatives industrielles*, Ed. de l'Entrepreneur, Paris, 1993

[161]Code Permanent Environnement et Nuisances, Ed. législatives

Table 1.12 Comparison of different sound levels.

Possibility of conversation	Auditory sensation	No. dB	Inside noise	Outside noise	Vehicle noise
	audibility threshold	0	acoustic laboratory		
	unusual silence	5	acoustic laboratory recording		
		10	studio		
	very quiet	15		gentle rustling of a leaf	
whispered voice		20	radio studio	quiet garden	
		25	quiet conversation at 1.50 m		
	quiet	30	apartment in quiet neighbourhood		
		35			sailing boat
normal voice	fairly quiet	40	calm office in quiet neighbourhood		
		45	normal apartment	minimal street noise during the day	1st class cruise ship
		50	quiet restaurant	very quiet street	smooth car
	everyday noise	60	department store, normal conversation chamber music	residential street	motor boat
fairly loud		65	noisy apartment		private car on road
	noisy but bearable	70	noisy restaurant, music	significant traffic	modern sleeping car
		75	typing office, standard factory		subway on tyres
difficult	difficult to bear	85	very loud radio, turning workshop	heavy traffic at 1 m	subway in movement, horns
		90			TGV
		95	forging workshop	street with heavy traffic	propeller aircraft
		100	band mill	pneumatic drill at less than 5 m	motorbike w/o silencer at 2 m, train wagon

(Continued)

Table 1.12 Continued.

Possibility of conversation	Auditory sensation	No. dB	Inside noise	Outside noise	Vehicle noise
need to shout	very difficult – to bear –	105	planing machine		subway (inside old wagons)
		110	sheet metal factory	clinching at 10 m	train passing in station
	pain threshold	120	motor test cell		aircraft motor at several metres
impossible		130	drop hammer		
	requires special protection	140	turbo-jet test cell		

According to INRETS and the OECD, depreciation of housing due to transport noise is valued at nearly 0.5 billion euro a year.

Reduction at source is achieved by interlaying absorbent materials in transmission bodies or under vibrating tools, reducing rotation speeds or the drop height of materials, and silencers on exhaust pipes. Noise propagation can also be brought down thanks to absorbent materials that reduce reverberation, particularly on the ground (e.g. flooring made from textiles or sound-absorbing plastic). In a sound-free field, the sound level drops by 6 dB (A) for each doubling of the distance from source. Screens can also be used. In buildings, technical solutions include floating floors on rock wool, acoustic linings made of plaster and rock wool, absorbent products on ceilings, acoustic double glazing and acoustic air inlets.

1.3.2 Degradation of ecosystems and landscapes

Mining of sand and gravel represented 17 million tonnes in 1950, 230 million tonnes in 1980, and over 400 million tonnes today[162], taken from 3000 sites. Granular materials (grains smaller than 80 mm) can be of alluvial origin (40% and dropping in France), or produced by crushing limestone or eruptive rock. Although there is currently no shortage, there are few remaining sites where mining causes only limited damage (vis-à-vis ecosystems like alluvial valleys, or landscapes). Alluvial gravel in the low-flow channel of rivers, which acts as a filter for the water table and supports fish spawning grounds, must be preserved.

For this reason, authorities encourage using substitutions in place of solid rock. In order to restore quarries at the end of mining, companies must plan storage of

---

[162]Éric Labouze, *Bâtir avec l'environnement, enjeux écologiques et initiatives industrielles*, Ed. de l'Entrepreneur, Paris, 1993 et www.lafarge.fr, 2008

covering material (e.g. topsoil), level out and clean the terrain, and possibly restore the original vegetation. Protecting the quality of water around the site also comes under the operators' specified responsibilities.

Integrating buildings into the landscape is a major aspect of architectural design. Computer-assisted design tools allow architects to visualize the project in its future environment using a scanned photo of the site. This can be used to carry out satisfaction surveys and compare different options.

Other potential nuisances caused by buildings include encroaching on the "right to sunlight" (shade from neighbouring buildings), deflecting the wind (draughts that affect comfort in the street and other areas close to high buildings), and increasing the temperature in highly urbanized zones, sometimes by several degrees.

1.3.3 Land use

Using land, in particular natural land, can be compared to consuming a resource that can become locally scarce. Artificial land takes up close to 5% of the French territory[163], with significant variations: from 0.4% in Lozère to 90% in Paris (15% in the North and Rhône areas)[164]. The priority should be to reuse areas that are already urbanized and maintain existing trees. Currently, each of the world's inhabitants has on average 0.26 hectares of land to satisfy his or her food requirements, which is insufficient: one French person, for example, has 0.8 hectares. Some experts predict a 20% reduction in this surface area by 2010, due to demographic growth[165]. In France, artificial land grew by around 5% from 1990 to 2000, mainly in Ile de France (Paris region), western Brittany, the plain of Alsace, the Rhodanian corridor and the Mediterranean coast.

1.3.4 Microclimate perturbations

A building can reduce the comfort of nearby outside areas by deflecting the wind (especially in the case of high buildings and narrow streets). It can also encroach on the "right to sunlight" of neighbouring buildings.

1.3.5 Odours

The olfactory mucosa that lines the inside of the nasal cavity is covered with very fine hairs submerged in a layer of mucus and implanted in cells that are sensitive to odour and linked to the olfactory nerve.

The most unpleasant-smelling products are nitrogen products and amines (e.g. methylamine, smells 8000 times stronger than ammonia, present in gas from quartering), aldehydes (butyraldehyde, with a rancid odour), organic acids (acetic, butyric and valeric acid, strong, pungent odour), organic solvents (in paints) and sulphur products (H_2S, 2500 times stronger smelling than ammonia, and mercaptans, which smell like rotten cabbage).

---

[163] Les données de l'environnement, IFEN, No. 101, March 2005
[164] Les données de l'Institut Français de l'Environnement, N°80, January–February 2003
[165] François Ramade, *Écologie appliquée*, Ediscience, Paris, 1995

Prevention solutions include designing leak-tight installations (restricting the length of pipes, the number of connections, the circulation of residual water in the open air, etc.) or modifying procedures (e.g. depressurizing the building, more frequent washing, ventilation, liming or precipitated sulphur). In addition, gas emission can be treated: condensation, incineration (thermal or catalytic), adsorption, ozonization, biofiltration (on a peat bed), dilution and the use of masking agents.

1.4 IN THE BUILDING ITSELF

The Anglo-Saxons introduced the concept of "sick building syndrome". Symptoms include absenteeism in tertiary buildings, health issues (headaches, fatigue, respiratory infections), and complaints from occupants – all supposedly induced by the building and/or its equipment (e.g. insufficient ventilation, badly designed or badly maintained air conditioning, materials and coverings that release pollutants, etc.). The links between the ailments observed and the pollutants emitted are not clearly established in this general concept.

Some pathogenic factors have however been isolated: asbestos, lead, some allergens, some volatile organic compounds, etc. Here we present some of the notions of air quality, water quality, and other health-related aspects (electromagnetic fields, comfort and risks).

1.4.1 Air quality and health

Air quality is primarily dependent on outside air (cf. paragraph 1.2.1). When choosing a construction site, it is important to examine any potential emission sources (factories, roads, etc.) and determine the direction of dominant winds. Some activities that take place inside buildings may however cause additional pollution.

The approximate composition of "clean", dry air at sea level is as follows: 78% nitrogen (N_2), 21% oxygen, 1% argon and 0.03% carbonic gas[166]. The other constituents (hydrogen, solid particles, rare gases, etc.) are present in lower quantities. CO_2 emissions of anthropogenic origin contain a 0.035% concentration of CO_2.

"Sick building syndrome", which is often of multifactorial origin, designates a set of symptoms (ocular, respiratory, cutaneous, stress) whose cause cannot be clearly identified. In France, the syndrome affects a reported 30% of new, air-conditioned buildings and between 10% to 30% of their occupants[167]. The causes include inadequate maintenance (badly cleaned filters, germs in pipes, etc.) and unsuitable design (e.g. damp problem).

Cooking with gas introduces NO_x. Water heaters and stoves that are badly adjusted give off CO, causing 300 deaths per year and around 8000 hospitalisations, often requiring the use of a hyperbaric chamber. Haemoglobin associates with CO in preference to oxygen and this fixation is irreversible. For a CO concentration of 800 ppm in the air, 50% of haemoglobin is transformed into carboxyhemoglobin,

---

[166] ASHRAE Handbook, *Fundamentals, chap. 11: air contaminants*, ASHRAE, Atlanta, 1993
[167] Éric Labouze, *Bâtir avec l'environnement, enjeux écologiques et initiatives industrielles*, Ed. de l'Entrepreneur, Paris, 1993

which reduces cellular oxygenation and attacks the central nervous system. Low doses of CO (urban air pollution for example) can cause the premature death of people with weak hearts. Moderate intoxication produces vague symptoms (e.g. headaches, dizziness, nausea), which makes diagnosis difficult. The toxicity threshold for prolonged exposure is a concentration of 0.1% volume. CO emissions have dropped considerably over recent years (5.2 million tonnes in 2006, i.e. a 70% reduction compared to 1973)[168]. CO is mainly released in the residential and tertiary sectors (33% of the total), industry (32%, of which 29% from metallurgy of ferrous metals) and road transport (24%).

Do-it-yourself home improvements also produce emissions: solvents (some of which damage the nervous system), pigments, welding (volatile lead), stripping (metals like lead, cadmium and chrome). Numerous aerosols used to clean ovens and windows, along with insecticides and house sprays give off harmful compounds like benzene. Metals accumulate in the body: lead in particular in the nervous system, cadmium in the kidneys, chrome in the lungs. Four glycol ethers were banned from sale to the general public in 1997 in concentrations higher than 0.5% (in paint, ink, varnish and cosmetic products). These products impact on reproduction. This regulation does not concern professionals, who can be exposed to doses higher than the reference dosage, which is obtained from animal experiments after applying a general safety factor. Toluene content of over 0.1% is banned in paint[169].

Tobacco smoke contains benzene, formaldehydes and dioxins. Formaldehyde is also emitted by some materials like pressed wood, some glues and foams (Urea-formaldehyde) and some carpets. Impacts on health are numerous (e.g. irritation of the eyes, headaches, breathing problems, sleep disturbance). The International Cancer Research Centre classes formaldehyde into group 2A of agents "probably carcinogenic to humans".

When vehicles are started up in garages joined onto housing, benzene (carcinogenic[170]) and other volatile organic compounds can enter living areas through doors (housing is often depressurized by the ventilation system). Similarly, air quality in housing is improved if the kitchen is separated from the other rooms.

Wood treatment products can irritate the skin, eyes and respiratory tract. They can also provoke head pain, dizziness and vomiting. Some doctors recommend using products free from lindane, pentachlorophenol and other organochlorines. Some metal-based products (e.g. chrome, copper and arsenic) are at times discouraged. Alternatives include oil bath treatments, an anhydride mix (renders wood water-repellent) and rectification (thermal wood-hardening treatment). Some manufacturers using these methods guarantee their woods for 30 years, without any notable incidence in terms of toxicity or eco-toxicity.

Fibres can be emitted from some insulating material, ventilation shafts and suspended ceilings. Over 70 varieties of artificial mineral fibre exist, some of which are classed as irritating or carcinogenic agents. Fibreglass contains larger fibres than rock

---

[168] CITEPA, Emissions in the air – national annual data, www.citepa.org, May 2008

[169] Official Journal of the European Union of 29-05-2007, annex XVII, Restrictions on the manufacture, placing on the market and use of certain dangerous substances, preparations and articles

[170] S. and P. Déoux, *L'écologie c'est la santé*, Ed. Frison-Roche, Paris, 1993

wool. The finer the fibres are, the deeper they penetrate into the organism. They can cause breathing and skin problems, and in 1988 the International Cancer Research Centre classed them in group 2B, "possibly carcinogenic to humans". Biopersistence is linked to the lifespan of a fibre in the organism, which depends on physiological mechanisms similar to dissolution. New fibres have been designed to degrade rapidly in the organism once inhaled, e.g. the half-life of glass wool is now 10 days. For this reason, since 2001 modern mineral wools have been classed in group 3, "unclassifiable as to carcinogenicity in humans". They remain highly irritating, however. It is fairly easy to ensure that these fibres remain confined using impermeable backing to avoid risks for occupants (risks are then restricted to manufacture, installation and removal).

Asbestos is a set of naturally fibrous metamorphic rocks: white asbestos or chrysotile, amphibole, blue asbestos or crocidolite, brown asbestos or amosite, tremolite, actinolite and anthopyllite. All forms of asbestos were banned in buildings in France on 1 January 1997, following a press campaign. As early as 1906, work inspectors described cases of pulmonary fibrosis linked to asbestos. The link between exposure to asbestos and lung cancer was established in 1955 thanks to an epidemiological survey carried out by Richard Doll in Great Britain and confirmed by numerous other studies. Only amphibole and crocidolite fibres were banned in 1988 in France, although other countries were stricter: Switzerland and Denmark banned asbestos in 1986. In France, the Ministry of Health estimates at least 600 deaths per year following exposure to asbestos, with a 25% progression every 3 years. Another source gives the figure of 3000 deaths per year[172]. Diagnosis and rehabilitation procedures exist following work done by the Institut National de Recherche et de Sécurité and the Comité Permanent Amiante. As an illustration, removing asbestos from the Montparnasse Tower (in 2005) cost 240 million euro.

Table 1.13 Asbestos substitution fibres.[171]

Toxicity estimated higher than asbestos
Fibres of silicon carbine, magnesium sulphate, calcium sulphate and glass
Toxicity estimated similar to asbestos
Fibres of vermiculite containing asbestos, wollastonite, attapulgite, aramid and phosphate
Toxicity lower than asbestos but potentially carcinogenic
Fibres of refractory ceramic, carbon, graphite, and all fibres with a diameter of under 3 μm
Toxicity unknown
Fibres of polymide resins or bismaleimides

In some regions with a granite subsoil or one rich in uranium or thorium, (in France the Massif Central, Vosges, Loire, Corsica and Brittany), infiltrations of radon from the subsoil make it necessary to increase ventilation to evacuate the pollutant (in particular isotope 222, and isotopes 219 and 220 whose lifespans are very short

[171]Franck Karg and Christophe Zeilas, *Les bâtiments pollués par l'amiante ou d'autres fibres à potentiel toxique*, Environnement et Technique, No. 255, April 2006
[172]Association nationale de défense des victimes de l'amiante, ANDEVA, 3 rue Sainte Félicité 75015 Paris

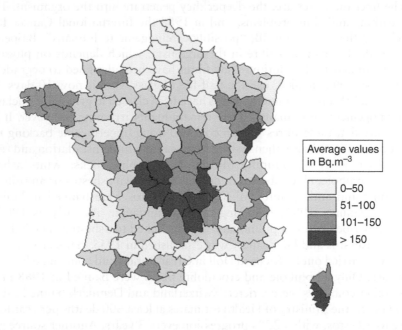

Figure 1.7 Map of volume activity of radon in habitations, 1982 to 2000, source: Institut de Radioprotection et de Sûreté Nucléaire (IRSN).

and therefore play a reduced role), and create a suitable design, e.g. proofing the ground-building interface, ventilating the basement or crawl space, creating positive pressure on the premises or depressurizing the crawl space. Radon, which is 8 times heavier than air, accumulates in low areas. The threat to health in fact comes from the derivatives that result from the disintegration of radon (especially polonium) and coat the pulmonary airways, whereas radon, a chemically inactive gas, is rapidly exhaled.

A 1995 study by the IPSN (French institute for radioprotection and nuclear safety) on uranium miners established a linear relationship between cumulated exposure to radon and an increase in the relative risk of death from lung cancer. An upper limit of 150 Becquerel/m^3 is viewed as acceptable in France (it is stricter in the Netherlands: 20 Becquerel/m^3), with the reserves indicated in the paragraph above concerning low doses of radioactivity. In 1990, the EC recommended intervention above 400 Becquerel/m^3 for existing buildings and 200 Becquerel/m^3 for new buildings. These standards have been applied in France since 2002. An order dating from July 2004 obliges employers and owners or operators of places open to the public (in priority teaching, health and penitentiary establishments) in 31 *départements* to measure the volume activity of radon. A measurement campaign carried out by the CEA (French Alternative Energies and Atomic Energy Commission) in Ile de France and 35 *départements* showed that these levels are only exceeded in respectively 1.5% and 5% of cases[173]. According to the IPSN, the level of 1000 Becquerel/m^3 is exceeded in 60,000

---

[173]Plan Construction et Architecture, call for propositions 1993: Experimental buildings of high environmental quality

dwellings (400 Becquerel/m^3 in 120,000). In the USA, from 10% to 15% cases of lung cancer in non-smokers[174] can be attributed to radon. Similar estimates attribute the death of about 2000 people in France to radon.

Inhaling crystalline silica is also associated with lung cancer[175].

The manufacture of paint containing lead (ceruse, lead hydrogen carbonate) has been banned since 1917, and its sale since 1948. Above a certain threshold, the ingestion or inhalation of lead is toxic: it causes reversible ailments (e.g. anaemia, digestion problems) and irreversible ones (attacks to the nervous system). Suitable methods exist to rehabilitate old housing (e.g. gentle stripping to limit dust, encapsulation, etc.).

Soft PVC, used in floor coverings for example, contains a significant proportion of phtalates (usually 35% to 40%, sometimes over 50%), which are emitted into the air when the products are used. Some of these phtalates present health risks[176], and six are banned in concentrations above 0.1% in mass of plastic matter in toys and infant care articles: bis(2-ethylhexyl) phthalate (DEHP), dibutyl phtalate (DBP), butyl benzyl phtalate (BBP), diisononyl phthalate (DINP), Diisodecyl phthalate (DIDP), and di(n-octyl) phthalate (DNOP)[177]. However, these substances are not regulated in flexible flooring, including in kindergartens and baby wards, where young children could come into contact with them. Stabilizers are added to prevent products degenerating through heat or light, for example lead-based products (tubes, profiles and cables), cadmium (chassis and windows, no longer used in Europe but in imported products) calcium and zinc, barium and zinc, organotins (roof coverings and rigid construction panels).

Inhaling biocontaminants can cause rhinitis, allergies, and in extreme cases, respiratory and lung infections, depending on the condition of the people exposed and the type of micro-organism. Mites are microscopic insects from the arthropod family and come under the arachnida class (4 pairs of legs) of the dermatophagoide type[178]. They proliferate in damp conditions (in particular unaired beds), and can cause allergies (e.g. asthma) via their droppings. Ventilating premises can reduce risks, but sometimes action is required on the allergen source.

If ventilation is insufficient, or if premises are unheated, the relative air humidity can be high and condensation can form on the coldest walls. If the relative humidity of a wall remains above 70% for long enough, then mould can appear. Conversely, in heated, scantly occupied premises, air is very dry because cold winter air introduced by ventilation contains little water vapour. Air can then be lightly humidified to reach a relative humidity level of between 40% and 60%. However, care should be taken not to introduce risks, linked to legionella.

---

[174]Éric Labouze, *Bâtir avec l'environnement, enjeux écologiques et initiatives industrielles*, Ed. de l'Entrepreneur, Paris, 1993

[175]Environmental Health Information Bulletin, published by the Quebec public health network – Volume 4 – No 3 – May–June 1993

[176]Commission of the European Communities, *Green Paper – Environmental issues of PVC*, July 2000

[177]Official Journal of the European Union of 29-05-2007, annex XVII, Restrictions on the manufacture, placing on the market and use of certain dangerous substances, preparations and articles

[178]Ministry of Health, Ministry of Equipment and Housing, Building and Health, December 1999

Some air conditioning systems or hot water production/supply systems present bacteriological risks because of unsuitable design (i.e. temperature favourable to micro-organisms). In July 1976, 221 cases of lung infection were identified at an American legion congress. The bacteria identified went on to be named Legionnella pneumophila. A benign form of legionella exists, similar to influenza syndrome, and a serious form, known as Legionnaire's Disease, which usually appears in fragile people (old or immuno-suppressed). The responsible germ can be found in all natural or artificial aquatic environments, in particular sanitary installations (i.e. showers, taps, hot water tanks and pipes), and air-conditioning units (water-cooling towers). The bacterium can be inhaled close to cooling towers or while taking a shower. It can be fatal if it reaches the lungs, and causes around one hundred deaths per year in France. The French by-law dated 27 April 1999 concerns operators and owners of installations that include water-cooling towers.

The Danish label on inside air quality evaluates the impact of compounds using chamber measurements of VOC, odour, asbestos, fibre, etc.[179]. The Blue Angel floor covering label applies a phtalate content threshold value of 0.1% in mass. This label has been granted to nearly 4000 products[180]. The association Natureplus also has a label[181]. This European eco label currently applies to paint (emissions below 30 gr VOC per litre), varnish, floor coverings (reduced content of heavy metals) and energy-saving light bulbs (mercury content limited to 4 mg per bulb compared to the standard 100 mg). The NF Environment standard applies to paint and varnish (VOC emissions under 100 g per litre of product). Regular paint contains around 50% solvent (350 g/l on average for traditional paint[182]), which ends up being emitted into the air, whereas emissions are reduced to 10% of the total content for some water-phase paint (190 g/l for acrylic dispersion type, 42 g/l for alkyd emulsion type[181]). In addition, some solvents are biodegradable (alkyd resins from soya for example, which are often mixed with other compounds). The environment and health declaration lists, based on a lifecycle assessment, are presented in the following chapters.

The table 1.14 presents an example of classification based on the characteristics of materials in terms of VOC, mould, bacteria and radioactivity[183]. VOC concentration is measured in an atmosphere of 23°C, with 50% humidity and an air speed of 0.2 m/s, which corresponds to usage conditions, the size of the sample and test cell being fixed by protocol (studies carried out in different standardization committees). Test standards currently used (ISO 16000 series) are not optimized to measure phtalates, since these compounds are generally on the boundary between volatile and semi-volatile compounds: longer testing would be preferable to identify low but persistent emissions.

The activity of building materials is measured in Bq/kg for the following three compounds: ^{226}Ra, ^{232}Th and ^{40}K. An excessive dose of gamma radioactivity is then evaluated in mSv by calculation, taking into account a product user scenario.

---

[179] www.esic.org

[180] www.blauer-engel.de

[181] www.natureplus.org

[182] Jean-Claude Laout, Formulation des peintures, mise en œuvre des polymères utilisés, Techniques de l'Ingénieur J2-272

[183] CSTB, Caractéristiques environnementales et sanitaires des produits de construction, Examination procedure of a CESAT case, July 2003

Table 1.14 Classification of materials regarding health.

Volatile organic compounds[184]	Radioactivity[185]
C– if concentration after 3 days >5 mg/m^3 or if concentration after 28 days >200 μg/m^3 or if concentration in carcinogenic compounds >1/10 of the concentration after 3 days C if the above thresholds are respected C+ if emission <50% of above thresholds	R– if activity > 100 Bq/kg for radium or thorium, or >1000 Bq/kg for potassium R if the above thresholds are respected R+ if emissions are very low
Bacteria	**Growth of fungus and mould[186]**
B– if vulnerable (bacterial growth on clean cell culture dish) B if inert (no bacteria survive on clean cell culture dish and bacteria present on soiled product) B+ if bacteriostatic (bacterial survival >50% on soiled product) B++ if limited bactericidal (bacterial survival <50% on soiled product) B+++ if bactericidal (no bacteria survive on soiled product)	F– if vulnerable (e.g. duck feathers), F– – if the biomass on the clean product is around 10 times lower than that of the soiled product, F– – – if the biomass on the clean product is equal or above that of the soiled product F if inert (absence of fungal growth on clean cell culture dishes and presence on soiled dishes) F+ if fungistatic (absence of fungal growth on clean and soiled dishes)

Mould can cause infections (aspergillosis) or produce toxins some of which are carcinogenic. The question of the durability of bactericidal and fungicide properties remains unanswered.

Other aspects are important, e.g. the emission of dust and fibres, heavy metals, ozone and non-ionizing radiation[187].

1.4.2 Water quality and health

The hardness of water is an indicator of its concentration in $CaCO_3$ equivalent. A French degree (°f) is equal to 10 mg/l of $CaCO_3$. The hardness of "soft" water is below 20°f, and hardness of over 35°f corresponds to "hard" water. Hardness encourages a build-up of scale deposit and so microbial proliferation. Conversely, softness causes greater dissolution of metals. Hardness has little impact on health since its effects are contradictory (increase in kidney stones, prevention of cardiovascular disease, but these possibilities are still controversial).

The massive use of fertilizer in agriculture coupled with intensive farming has contributed to increasing the nitrate levels in most water supply networks. In some

---

[184]European Collaborative Action Indoor Air Quality and its Impact on Man, Evaluation of VOC emissions from building products – solid flooring materials, European Commission, 1997
[185]Directorate General Environment, Nuclear Safety and Civil Protection, Radiological Protection Principles concerning the Natural Radioactivity of Building Materials, European Commission, 1999
[186]AFNOR, Standard NF-EN-ISO 846, Plastiques: évaluation de l'action des micro-organismes, August 1997
[187]Drs. S. and P. Déoux, *Le guide de l'habitat sain*, Medieco Editions, Andorre, April 2002

regions, levels have multiplied by five. Excessive ingestion of nitrates can have consequences, particularly on babies. Nitrates are metabolized into nitrites, which react with haemoglobin to form methaemoglobin, which is unsuitable for transporting oxygen[188]. In babies, this reaction is amplified due to low gastric acidity and because the methaemoblobin is not transformed into oxyhemoglobin, unlike in adults. In addition, the nitrites formed after absorbing nitrates in turn produce nitrosamines, most types of which are carcinogenic.

Pesticides used in agriculture generally comprise 46% herbicide, 31% insecticide and 18% fungicide. Water suitable for drinking should not contain over 0.1 µg/l per pesticide, and total pesticides should not exceed 0.5 µg/l. Like heavy metals, these products can be concentrated in the food chain. Numerous types of pesticide exist (over 35,000 products). Organochlorines break down with difficulty, both in the environment and in the human body. DDT and all organochlorines except for lindane are banned in France. Risks linked to pesticides include cancer and chronic neurotoxicity (Parkinson's Disease). It is possible that agent orange (a defoliant used in Vietnam) affects reproduction, that organosphosphate affects respiration, and that organochlorine affects the immune system.

The presence of heavy metals in water or in the food chain also has significant consequences (some vegetables concentrate these metals when low quantities are present in irrigation water). Lead pipes have been banned since 1978 for water supply, and since 1995[189] in buildings. Consuming water that has remained in lead pipes is not recommended. Lead can cause numerous ailments, even at very low concentrations, such as anaemia, neurotoxic effects, kidney problems, and alterations to reproduction. The European directive of 03/11/98 on the quality of water for human consumption has required lead concentration of under 25 µg/l since the end of 2003 and 10 µg/l from the end of 2013. The first threshold calls for treating water using filmogen or corrective; the second will necessitate replacing lead pipe systems (mainly in buildings built prior to 1949). Brass taps also release lead (brass contains 8% lead) in non-negligible quantities during the first months of usage. Manufacturers envisage stocking such taps for several months before putting them on sale. In the long term, drinking water concentration limits could also be imposed for other substances: boron, arsenic, polycyclic aromatic hyrocarbons, chlorine solvents, nickel, antimony, copper, benzene, bromates, and trihalomethanes including chloroform.

The acidity of water makes soluble some of the aluminium contained in rock, which can then reach aquifers. Aluminium salts can also be used in water treatment plants to eliminate some minerals and organic matter. The European Commission has set an aluminium concentration threshold of 0.2 mg/l after treatment. Possible consequences of ingesting aluminium are osteoporosis (affecting the bones) and Alzheimer's Disease. This latter disease is the fourth cause of death in developed countries, following heart disease, cancer and cerebrovascular accidents. In the USA, it affects 10% to 15% of the over-65s.

Some mineral water with high fluoride content can cause fluorosis (which attacks the bones through lack of calcium). Packaging used for mineral water can affect its

---

[188] S. and P. Déoux, *L'écologie c'est la santé*, Ed. Frison-Roche, Paris, 1993
[189] French Act No. 95-363 of 7 April 1995

quality: PVC bottles that are kept for too long can be carcinogenic. Glass bottles contain lead. In some regions, radon is present in the water supply, bringing a risk of cancer.

Regarding the risk of legionellosis from the hot water supply, several prevention and treatment methods exist. Heat sterilization involves raising the temperature of all reservoirs and circuits for a given length of time each day to eliminate bacteria (e.g. one hour at 60°C). Chlorine treatment is also possible. Maintaining installations, in particular regular descaling, is indispensable, since the bacteria mainly feed off scale deposits.

The bylaw of 30 November 2005 concerns domestic hot water installations, and raises two contradictory concerns: in order to reduce risk from scalding, water temperature should not exceed 50°C at withdrawal points in rooms used for washing (60°C in other rooms, and up to 90°C in collective kitchens and laundries, with appropriate signs); but to reduce the risk of legionellosis, the temperature must be above 50°C throughout the supply circuit, and above 55°C when it leaves equipment; when stored, the temperature in the water reserve must be raised to 60°C every day for at least an hour (or at least 4 minutes at 65°C, or 2 minutes at 70°C).

As it leaves the treatment plant, a dose of chlorine is added to water to ensure its bactericidal potential up to the tap and to avoid recontamination in the water supply system. The taste of chlorine can be attenuated by keeping water in a refrigerator: when the temperature drops, the molecular agitation slows down and the chlorine molecules are less dispersed. Chlorinating water that is rich in organic matter produces halomethanes (like chloroform), reputed to be carcinogenic, so that a compromise needs to be found between protecting from bacteria and limiting the risk of cancer.

The insides of water pipes are colonized by a fine film of non-pathogenic bacteria called "biofilm". This film can act as food for pathogenic germs, and it is useful to reduce it by bringing down the level of dissolved organic carbon. Springs or private wells require adequate maintenance to avoid micro-biological pollution (i.e. distance from livestock, surface proofing and regular purification using bleach).

Negative ions are increasingly present in some natural sites (e.g. close to mountain waterfalls). They are much rarer in towns, due to the high content of dust, aerosols and various micro-organisms in the air. Close to the ground, an average of one molecule in 10^{16} is ionized[190], i.e. a concentration of around 2 to 3000 ions per cm^3 in a non-polluted outside atmosphere. Cosmic radiation produces two pairs of ions per cm^3 and per second at sea level, and 3 pairs at 2000 m altitude. The radioactivity of rock produces from 2 pairs (sandy terrains) to 100 pairs (some granite terrains). Radon produces five. Vegetation produces negative ions (chlorophyllous function, emission of electrons by very pointed conifer leaves). Television and computer screens, on the other hand, produce positive ions. Negative ions have a germicide effect[191], and some doctors recommend producing them in buildings.

---

[190]Prof. J. Breton, *Certificat international d'écologie humaine*, Lecture at Bordeaux 1 University, 1986

[191]Jacques et Micheline Breton, *Climat et Santé*, Report by the Dijon Faculty of Medicine, 1994

1.4.3 Electromagnetic fields

The Earth's magnetic field is around 0.5 Gauss. An additional magnetic field can exist around a geological fault, of around 10^{-9} Gauss. For artificial magnetic fields, a 2.5 mG limit is recommended for a frequency of 50 Hz, meaning that we should keep several tens of cm away from electric equipment, and a little further from the back of old computers and televisions with cathode ray tubes, from which the sweep frequency is 20,000 Hz or more (but flat screens are now standard). This threshold is 200 times lower than the Earth's magnetic field, but the fields are pulsed, whereas the natural field is constant.

Some doctors recommend avoiding chronic exposure to pulsed electromagnetic fields (e.g. electric blankets, cathode tubes, etc.) and, for example, keeping radio alarm clocks at least 70 cm away from the bed. Building close to high-voltage electricity lines is also not recommended. The effects of these rays, which are difficult to prove, are reported to be headaches, fatigue and insomnia at extremely low frequencies (50 Hz), and a risk of miscarriage at very low frequencies (50 kHz, close to cathode screens). Electrostatic pollution (e.g. synthetic carpet) causes a higher concentration of dust around people and a risk of bacterial contamination.

For lighting, halogen bulbs that are not protected by a double glass envelope give out ultraviolet rays, some of which are harmful to health. Fluorescent tubes, with a pulsation of 100 Hz, can cause headaches and their dominant blue/violet light can accelerate ageing of the retina.

1.4.4 Sources of discomfort

The main sources of discomfort in buildings may be hygrothermic (overheating in summer under a roof or in space with a lot of glass, drafts in a space with too much ventilation or unsuitable air conditioning, air that is too dry or humid); visual (over-bright or dark); olfactory (insufficient ventilation); or acoustic (inadequate protection against outside noise, neighbouring housing or traffic).

The combined impact of these factors can be significant. For example, the influence of colour on heat sensations is well known: heat-sensitive Ruffini corpuscles are located deep within the skin and are only accessible to very long-wave light rays (red and yellow); cold-sensitive Krause corpuscles are on the other hand more superficial. Psychological impacts also exist, e.g. the longest wavelengths shrink space. Asthma sufferers thus tend to prefer green and blue shades, which expand the perception of space. The interaction between heat, hygrometry and air speed is also well known, and can be represented in a diagram like the one below, where the dry temperature is easting and absolute humidity northing (i.e. specific humidity in g of water per kg of dry air). We can then draw curves showing relative iso-humidity (100% corresponds to the dew curb). The surfaces A and B represent comfort zones. Surface B is larger and corresponds to a higher air speed (e.g. presence of a fan or draught), which increases heat transfer by convection and thus makes it easier to bear high temperatures.

In fact, thermal comfort integrates different modes of transfer between the body and its environment[192]: convection (exchange with air), radiation (exchange with the

---

[192]Françoise Thellier, *L'Homme et de son environnement Thermique. Modélisation.* Paul Sabatier University, Toulouse, 1999, 60p

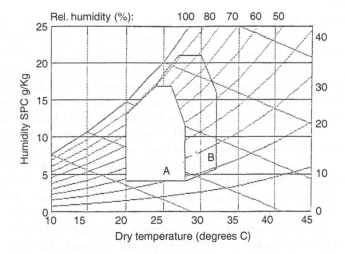

Figure 1.8 Example of a diagram representing comfort zones according to temperature, humidity and air speed.[193,194]

walls of a premises), and to a lesser extent, conduction (exchange with the ground). Outside the comfort zone, different regulatory mechanisms set in (sweating, increase in skin temperature through the blood – the skin reddens, shivering, clothing). The relevant parameter for thermal comfort is therefore not air temperature, and so the "operative" temperature obtained through convective and radiative exchange is often used. For slow air speed (0.1 m/s), this temperature is the average between air temperature and average wall temperature (weighted by surface). A more general method evaluates an average voting percentage depending on people's activity, their clothes, the temperature of the air and walls, the air speed and the humidity[195,196]. This evaluation results from tests during which 1300 people grade atmospheres from +3 (hot), +2 (warm), +1 (slightly warm), 0 (neutral), −1 (slightly cool), −2 (cool), and −3 (cold). It is recommended that premises be designed so that the average vote percentage is between −0.5 and + 0.5 (which statistically corresponds to less than 10% of dissatisfied people).

A great deal of publications deal with hygrothermic, visual, acoustic and olfactory comfort, and they are thus are not covered in more detail here.

[193] B. Givoni, *L'homme, l'architecture et le climat*, ed. du Moniteur, Paris, 1978, 460p

[194] Dominique Campana, François Neirac and Gabriel Watremez, *Elaboration d'un logiciel sur micro-ordinateur pour l'aide à la conception en pays tropical sec*, Interministerial programme REXCOOP, Report by the Ecole des Mines de Paris, April 1995, 59p

[195] Fanger P.O. *Thermal Comfort*. McGraw Hill book compagny, New York, 1970

[196] Standard NF EN ISO 7730, Ergonomie des ambiances thermiques – Détermination analytique et interprétation du confort thermique par le calcul des indices PMV et PPD et par des critères de confort thermique local, 1985, revision 1994 then 2005

1.4.5   Risks

CO intoxication is the first cause of toxic death in France, with over 8000 cases per year and 300 deaths[197]. The absence of a chimney in some housing heated by electricity can increase this risk if the occupant decides to change energy (to fuel oil or gas).

Fire is also a major risk in buildings, and regulations make it obligatory to use fire-resistant materials. Those classed as M0 (non-combustible materials, European class A1) and M1 (materials very difficult to inflame, class A2) correspond to materials that do not generate toxic smoke. Some fires have burned down buildings open to the public that include a significant quantity of PVC without adequate protection, and these have drawn attention to this material, which can release hydrochloric acid and dioxin when it burns (e.g. the fire at a discotheque in Germany where 161 people died). The Syndicat National des Plastiques Alvéolaires[198] produces a brochure presenting ways of improving walls' resistance to fire (e.g. plastic materials inserted into sandwich panels, screen materials, flameproof facings), making them resistant to fire for around 20 to 30 minutes, even class M2 materials and compounds (i.e. difficult to inflame, European class B). Making buildings secure requires an overall approach that is not restricted to materials (e.g. suitable design, compartmentation, detection and extinction systems, prevention, maintenance, fire exits and evacuation plans, etc.).

Transporting energy brings risks, whether gas (e.g. explosion of gas pipeline or fire), oil products (110 road accidents per year, 4 oil tanker accidents affecting the French coastline in 15 years) or electricity.

In total, accidents in the home kill around 20,000 people a year, which is very high in comparison to 5000 deaths on the road and 1000 fatal accidents in the work place.

Other risks are linked to the building's actual environment: construction in flood zones, or in zones under threat of landslide, earthquake or subject to violent wind. Design rules exist to protect against climate (rain, snow, wind and hail) and earthquakes. The choice of site is a crucial factor.

1.5   SUMMARY OF IMPACTS

The overview presented above illustrates the serious nature of some issues, but also the progress made in several areas (ozone layer, acid rain, VOC emissions, etc.). The most significant improvements are generally linked to regulations following international agreements (e.g. the Montreal and Gothenburg Protocols), but voluntary agreements sometimes pave the way for efficient technical solutions and help regulations to move forward.

This impact analysis is simplified given that in reality a very complex chain of events takes place. For example, carbon gas emissions modify the optical properties of the atmosphere. As a result the climate is perturbed, with regional variations. Meteorological models are not yet capable of predicting the consequences of this global

---

[197] Ministry of health, Department of health, sub-department managing risks and environments, Bureau SD7C: Buildings, noise and the work place, values obtained by extrapolation of 2002 statistics for Ile de France, www.sante.gouv.fr, October 2004
[198] SNPA, 15 avenue du recteur Poincaré, 75016 Paris

perturbation at local level (storms, floods, temperature rise, etc.). The increase in temperature in some regions could then impact on health (e.g. wider spread of malaria), and affect biodiversity (some plants, trees in particular, will not survive if climate zones move too quickly). These chains of consequences cannot be evaluated on the base of current knowledge, which is why decisions on climate protection are based on the precaution principle: even if the consequences of a phenomenon are not precisely known, if they are potentially very serious, then it is preferable as a precaution to reduce the emissions causing the phenomenon.

The issue is also made more complex by the number of aspects considered, and possible contradictions between them. For example, a higher demand for comfort in the summer can involve using air conditioning thus leading to higher energy consumption and the consequent impacts. A multi-criteria approach is generally applied. Put simply, grades are attributed to each theme for the different variations proposed: in the previous example, the solution without air conditioning would have 0 for comfort and 5 for energy, and with air conditioning it would have 5 for comfort and 0 for energy. Weighting factors are defined for each aspect (e.g. 0.5 for comfort and 0.5 for energy), which makes it possible to give an overall grade. The multi-criteria approach is obviously more subtle than this simplified presentation, but the choice of weighting factors is still based on subjective considerations linked to context and decision-makers' priorities.

Another, more complex method is to take a synthesis approach to ironing out such contradictions. In the above example, it is often possible to adjust the thermal mass in a building so as to limit or even cancel out the need for air conditioning, e.g. using a heavy masonry slab to attenuate temperature amplitudes. The flooring would also store solar energy in winter and so reduce the heating load. The amount of energy needed to manufacture a slab is low compared to that used to heat a building (see chapter 3).

This solution therefore creates a synthesis between summer comfort and winter heat. However, it requires efficient thermal contact between the flooring and the atmosphere, so as to avoid for example putting a thick carpet on the masonry floor. This can be in contradiction with another aspect of comfort or quality of life: to be able to walk barefoot without having cold feed. The priority given to these different aspects influences the final decision made, which will necessarily be based on subjective choices. The degree of comfort required is subjective in itself and varies depending on the context: a century ago, an apartment heated to 15°C was considered comfortable; in 1973 the French "energy-saving" campaign recommended 19°C, whereas today temperatures of 22°C to 23°C are not uncommon.

Uncertainties on assessing impact and applying the precaution principle, along with the subjective choice of priorities regarding environmental aspects, mean that environmental quality can be defined in very different ways. Whatever the case, the seriousness of environmental issues is now indisputable and it is no longer reasonable to ignore the environmental consequences of the decisions we make. The building sector makes a significant contribution to these problems. As far as possible, it is therefore necessary to make use of decision-making tools and technologies for reducing environmental impacts in this sector.

Chapter 2

Environmental indicators

Indicators corresponding to most of the environmental themes presented in chapter 1 are used to assess the potential impacts of technical choices and help guide decisions. This chapter provides a definition of these indicators, which can initially be considered as performance indicators: a deeper analysis considers a much vaster chain of environmental effects (cf. conclusion of the preceding chapter). Some authors thus make a distinction between mid-point indicators and end-point indicators.

These overall performance indicators can be supplemented by partial indicators, which are easier to evaluate. For example, the consumption of non-renewable energy and water, the proportion of recycled materials used in a building, the proportion of waste sorted, etc.

It can also be worth looking at urban environmental indicators to determine the contribution a building makes to the environmental performance of the neighbourhood, or even the town, in which it is located.

2.1 POTENTIAL, OR MID-POINT INDICATORS

We have already seen how difficult it is to establish a link between emissions and actual impacts (e.g. between carbon emissions and the damaged caused by storms and floods). Given the current state of knowledge, it is worth applying a principle of precaution. This leads to the use of potential, or mid-point indicators. A potential impact will not necessarily occur, but an indicator is useful, since reducing potential impact reduces the risk of actual impact. The global warming potential (GWP) indicator comes into this category, as do acidification and eutrophication indicators, but for different reasons: the actual impact depends on the location of the emissions.

Emissions at the origin of the acid rain phenomenon only actually impact vegetation if their concentration is sufficiently high. The impact depends not just on the emissions introduced by the procedure or product studied, but also by what is known as the "background concentration", linked to all other emissions in the same region. The emissions evaluated by the indicator, which are only linked to the procedure or product studied, therefore only have a potential impact.

Regarding eutrophication, the impact is also linked to the concentration of pollutant, but it will depend more on the dilution of the products than on their quantity. Thus, a given quantity of pollutant will have a greater actual impact if it is emitted in a small lake than if it is released into the ocean. The consequences are, however, visible on the coastline, with the blue-green algae phenomenon.

Indicators are evaluated according to the "elementary flows", i.e. the quantities of substance released into the environment (without subsequent human modification) or taken from the environment (without prior human modification). Taken together, the elementary flows constitute the inventory of a product or procedure in the sense of a life cycle assessment. A characterization factor links each elementary flow to each indicator, e.g. for an indicator of the contribution to the greenhouse effect, the GWP of methane is 25 kg of CO_2 equivalent per kg of methane emitted. By giving $C_{i,j}$ the characterization factor of the i^e elementary flow (F_i, expressed in kg) the indicator I_j corresponding to the impact j is calculated by:

$$I_j = \sum_i C_{i,j} \cdot F_i \qquad (2.1)$$

For example, the global warming GWP indicator is expressed in weight of CO_2 equivalent: the relative contribution of gas emitted into the atmosphere depends on its optical properties. Since the substances that contribute to global warming do not necessarily have the same lifespan in the atmosphere as CO_2[199], the characterization factors depend on the duration of the analysis considered. As an illustration, the global warming indicator at 100 years is calculated by:

$$GWP_{100} = 1 \cdot F_1 + 25 \cdot F_2 + 298 \cdot F_3 + 22\,800 F_4 + \sum_{i>4} C_i \cdot F_i \qquad (2.2)$$

where F_1 to F_4 are respectively the quantities of CO_2, methane, N_2O, and SF_6 emitted. Flows of index i higher than 4 correspond to CFCs, HCFCs, HFCs, HCs, and halons etc.: the list of these substances and the corresponding values of coefficients C_i are given below.

The tables below show the factors used to undertake the aggregation of the environmental indicator assessment. These values reflect the current state of knowledge and could therefore change as knowledge progresses. Unless indicated otherwise, they are taken from a CML report (Guinée ed., 2001)[200].

2.1.1 Greenhouse effect

The global warming potential of different greenhouse gases evolves with our knowledge of the gases' optical properties and their lifespan in the atmosphere (IPCC, 1994[201]),

---

[199]The lifespan of CO_2 in the atmosphere is around 120 years, but only about ten years for methane.
[200]Guinée J. B., (final editor), Gorrée M., Heijungs R., Huppes G., Kleijn R., de Koning A., van Oers L., Wegener Sleeswijk A., Suh S., Udo de Haes H. A., de Bruijn H., van Duin R., Huijbregts M. A. J., Lindeijer E., Roorda A. A. H., Weidema B. P.: *Life cycle assessment; An operational guide to the ISO standards*; Ministry of Housing, Spatial Planning and Environment (VROM) and Centre of Environmental Science (CML), Den Haag and Leiden, Netherlands, 2001, 704 p.
[201]Intergovernmental Panel on Climate Change (IPCC), *Scientific assessment working group of IPCC, Radiative forcing of climate change*, World Meteorological Organization and United Nations Environment Programme, 1994, 28p

(IPCC, 2001[202]) and (IPCC, 2007[203]). The most recent values are given in table 2.1 for the timescales of 20, 100 and 500 years. The indicator is expressed in kg of CO_2 equivalent. A period of 100 years is considered in this analysis because it corresponds to the lifespan of CO_2, which is the main greenhouse gas in the building sector. However, some authors recommend considering a shorter timescale (Dessus et al., 2007[204]) to encourage reducing gas emissions with more short-term impacts, like methane.

The GWP values depend on the gases' optical properties (absorption factor of infra-red radiation emitted by the Earth, noted as "a", but also on how fast the gases degrade. The values are obtained from the integration over a certain time period (e.g. 100 years) of the product of "a" through concentration over time c(t), and given in CO_2 equivalent:

$$\text{GWP} = \int\limits_{100 \text{ years}} a \cdot c(t) \, dt / \int\limits_{100 \text{ years}} a_{CO_2} \cdot c_{CO_2}(t) \, dt \qquad (2.3)$$

Their uncertainty is estimated by the Intergovernmental Panel on Climate Change as $+/-35\%$, in terms of direct greenhouse effect. Our knowledge of the indirect greenhouse effect (i.e. effect on the number and size of cloud droplets, the lifespan of clouds and that of other greenhouse gases, etc.) is less precise.

We include CO_2 of non-fossil origin in our evaluation to cover the use of wood materials in construction (absorption of CO_2 during tree growth, emissions from landfill storage and incineration, and possibly energy recovery). Water vapour is not included in the indicator because on average water molecules only remain in the atmosphere for ten days.

2.1.2 Destruction of the stratospheric ozone layer

The characterization factors of gases that attack the stratospheric ozone layer are listed in table 2.2. These factors are called ODP (Ozone Depletion Potential). The indicator is expressed in kg of CFC-11 equivalent.

2.1.3 Acidification

The characterization factors of substances that contribute to acidification are listed in table 2.3. These factors are called AP (Acidification Potential). The indicator is

---

[202]IPCC, Houghton J. T., Ding Y., Griggs D. J., Noguer M., van der Linden P. J. and Xiaosu D.: *Climate Change 2001: The Scientific Basis*, IPCC, Intergovernmental Panel on Climate Change, Cambridge University Press, The Edinburgh Building Shaftesbury Road, Cambridge, UK, ISBN-13: 9780521014953, July 2001, 892p

[203]IPCC, Forster, P.M. (2007) Changes in Atmospheric Constituents and in Radiative Forcing, In: Solomon, S., D. Qin, M. Manning, Z. Chen, M. Marquis, K.B. Averyt, M. Tignor and H.L. Miller (ed.), Climate Change 2007: The Physical Science Basis. Contribution of Working Group I to the Fourth Assessment Report of the Intergovernmental Panel on Climate Change, Cambridge University Press, Cambridge, United Kingdom and New York, NY USA, 2007

[204]Benjamin Dessus, Hervé Le Treut and Bernard Laponche, Effet de serre, n'oublions pas le méthane, La Recherche, No. 417, March 2008

Table 2.1 Global warming potential of greenhouse gases.

Formula	Substance	GWP_{20}	GWP_{100}	GWP_{500}
CO_2	carbon dioxide	1	1	1
CH_4	methane	72	25	7.6
N_2O	dinitrogen oxide	289	298	153
SF_6	sulphur hexafluoride	16300	22800	32600
NF_3	nitrogen trifluoride	12300	17200	20700
CF_4	tetrafluoromethane	5210	7390	11200
C_2F_6	hexafluoroethane	8630	12200	18200
C_3F_8	octafluoropropane	6310	8830	12500
C_4F_{10}	perfluorobutane	6330	8860	12500
$c\text{-}C_4F_8$	octafluorocyclobutane	7310	10300	14700
C_5F_{12}	dodecafluoropentane	6510	8180	13300
C_6F_{14}	perfluorohexane	6600	9300	13300
CH_3Cl	chloromethane (HCC40)	45	13	4
CH_2Cl_2	dichloromethane	31	8.7	2.7
$CFCl_3$	CFC-11	6730	4750	1620
CF_2Cl_2	CFC-12	11000	10900	5200
CF_3Cl	CFC-13	10800	14400	16400
CHF_2Cl	HCFC-22	5160	1810	549
$C_2F_3Cl_3$	CFC-113	6540	6130	2700
$C_2F_4Cl_2$	CFC-114	8040	10000	8730
C_2F_5Cl	CFC-115	5310	7370	9990
$CHCl_2CF_3$	HCFC-123	273	77	24
$CHFClCF_3$	HCFC-124	2070	609	185
CHF_2CF_3	HFC-125	6350	3500	1100
CH_2FCF_3	HFC-134a	3830	1430	435
CH_3CFCl_2	HCFC-141b	2250	725	220
CH_3CF_2Cl	HCFC-142b	5490	2310	705
CH_3CF_3	HFC-143a	5890	4470	1590
CH_3CHF_2	HFC-152a	437	124	38
$C_3F_5HCl_2$	HCFC-225ca	429	122	37
$C_3F_5HCl_2$	HCFC-225cb	2030	595	181
C_3HF_7	HFC-227ea	5310	3220	1040
CHF_3	HFC-23	12000	14800	12200
$C_3H_2F_6$	HFC-236fa	8100	9810	7660
$CHF_2CH_2CF_3$	HFC-245fa	3380	1030	314
$C_2H_2F_2$	HFC-32	2330	675	205
$CF_3CH_2CF_2CH_3$	HFC-365mfc	2520	794	241
$CF_3CHFCHFCF_2CF_3$	HFC-43-10mee	4140	1640	500
CCl_4	HC-10	2700	1400	435
CH_3Br	bromomethane	17	5	1
CH_3CCl_3	HC-140a	506	146	45
CF_3Br	HALON-1301	8480	7140	2760
$CBrClF_2$	HALON-1211	4750	1890	575
$CBrF_2CBrF_2$	HALON-2402	3680	1640	503
CH_3OCH_3	dimethylether	1	1	$\ll 1$
CF_3OCHF_2	HFE 125	13800	14900	8490
CHF_2OCHF_2	HFE 134	12200	6320	1960
CH_3OCF_3	HFE 143a	2630	756	230
$CF_3CHClOCHF_2$	HCFE 235da2	1230	350	106

(Continued)

Table 2.1 Continued.

Formula	Substance	GWP_{20}	GWP_{100}	GWP_{500}
$CH_3OCF_2CHF_2$	HFE 245cb 2	2440	708	215
$CH_3CH_2OCHF_2$	HFE 245fa2	2280	659	200
$CHF_2CF_2OCH_3$	HFE 254cb2	1260	359	109
$CH_3OCF_2OCH_2CF_3$	HFE 347mcc3	1980	575	175
$CHF_2CF_2OCH_2CF_3$	HFE 347pcf2	1900	580	175
$CH_3OCF_2CF_2CHF_3$	HFE 356pcc3	386	110	33
$C_4F_9OCH_3$	HFE 7100	1040	297	90
$C_4F_9OC_2H_5$	HFE 7200	207	59	18
$CHF_2OCF_2OC_2F_4OCHF_2$	H-Galden 1040x	6320	1870	569
$CHF_2OCF_2OCHF_2$	H-G 10	8000	2800	860
$CHF_2OCF_2CF_2OCHF_2$	H-G 01	5100	1500	460

Table 2.2 Ozone destruction potential.

Formula	Substance	ODP
$CFCl_3$	CFC-11	1.0
CF_2Cl_2	CFC-12	0.82
$C_2F_3Cl_3$	CFC-113	0.9
$C_2F_4Cl_2$	CFC-114	0.85
C_2F_5Cl	CFC-115	0.4
$CHCl_2CF_3$	HCFC-123	0.012
$CHFClCF_3$	HCFC-124	0.026
CH_3CFCl_2	HCFC-141b	0.086
CH_3CF_2Cl	HCFC-142b	0.043
CHF_2Cl	HCFC-22	0.034
$CF_3CF_2CHCl_2$	HCFC-225ca	0.017
CF_2ClCF_2CHClF	HCFC-225cb	0.017
CCl_4	HC-10 (tetrachloromethane)	1.2
CH_3CCl_3	HC-140a (1,1,1-trichloroethane)	0.11
CF_3Br	HALON-1301	12
CF_2BrCl	HALON-1211	5.1
CBr_2F_2	HALON-1202	1.25
$C_2F_4Br_2$	HALON-2402	7
CHF_2Br	HALON (HBFC)-1201	1.4
$CHFBr-CF_3$	HALON (HBFC)-2401	0.25
$CHClBr-CF_3$	HALON (HBFC)-2311	0.14
CH_3Br	bromomethane	0.37
CH_3Cl	chloromethane (HCC-40)	0.02

expressed in kg of SO_2 equivalent, or sometimes in H^+ equivalent (1 kg SO_2 eq. = 32 kg H^+ eq.).

It may seem surprising that ammonia participates in acidification, since ammonium is a base. In fact, this gas decomposes and several chain reactions lead to the formation of nitric acid in the atmosphere.

Table 2.3 Acidification potential.

Formula	Substance	AP
SO_2	sulphur dioxide	1.0
SO_3	sulphur trioxide	0.8
NO	nitric oxide	1.07
NO_2	nitrogen dioxide	0.7
NO_x	nitrogen oxides	0.7
NH_3	ammonia	1.88
HCL	hydrochloric acid	0.88
HF	hydrofluoric acid	1.6
H_2S	hydrogen sulphide	1.88
HNO_3	nitric acid	0.51
H_3PO_4	phosphoric acid	0.98
H_2SO_4	sulphuric acid	0.65

2.1.4 Eutrophication

The characterization factors of substances that contribute to this phenomenon are listed in table 2.4. These factors are called EP (Eutrophication potential). The indicator is expressed in kg of PO_4^{3-} equivalent.

Table 2.4 Eutrophication potential.

Formula	Substance (emissions in water)	EP
NO	nitric oxide	0.2
NO_2	nitrogen dioxide	0.13
NO_x	nitrogen oxides	0.13
NO_3^-	nitrate	0.1
NH_4^+	ammonium	0.33
NH_3	ammonia	0.35
HNO_3	nitric acid	0.1
N	nitrogen	0.42
PO_4^{3-}	phosphate	1.0
H_3PO_4	phosphoric acid	0.97
P	phosphorus	3.06
P_2O_3	phosphorus trioxide	1.34
COD	Chemical oxygen demand	0.022

2.1.5 Winter SMOG

The characterization factors of substances that contribute to this phenomenon are listed in table 2.5. The indicator is expressed in kg of dust (Goedkoop et al., 1995)[205].

---

[205] M. Goedkoop et al., *The Eco-indicator 95, final report*, NOVEM, 1995

Table 2.5 Winter SMOG indicator.

Carbon	I
Dust	I
Iron dust	I
SO_2	I
Soot	I

2.1.6   Summer SMOG (photochemical ozone formation)

The characterization factors of substances concerned are listed in table 2.6 (Guinée ed., 2001)[206]. The indicator is expressed in kg of ethene (C_2H_4) equivalent.

2.2   CRITICAL VOLUME INDICATORS

This type of indicator is used to evaluate effects that depend on the concentration of pollutants whose noxiousness differs. In such cases we cannot simply add up the quantities of pollutants since their effect is not identical. Therefore, for each pollutant, a tolerable maximum concentration is defined C_m (kg/m^3), so that for example 95% of individuals in the environment considered are preserved. The critical volume is obtained by dividing the emissions by C_m.

For example, the aquatic ecotoxicity indicator is expressed as the sum of critical volumes (m^3 of polluted water) for the different pollutants. The more noxious a pollutant is, the weaker the C_m concentration (a weak concentration is sufficient to produce significant damage), and thus the higher the critical volume, for a given emission. Critical volumes can therefore be added together for different pollutants. The main limitation of this method is that it does not take into account the lifespan of pollutants, linked to their degradation in different environments.

More precise methods have therefore been devised: indicators are derived from models that make it possible to follow the transportation of pollutants between different ecological compartments (air, river water, soil, groundwater, etc.), their possible (bio)degradation, their transferral into the air, water and food, the doses received by the population concerned, and the effects (disease, victims). However, these models require more data, and these data are only available for a small number of substances (180, and even less, 46, for the EUSE model, which takes background concentration into account). For the time being, models do not consider interactions between different substances.

---

[206] Guinée J. B., (final editor), Gorrée M., Heijungs R., Huppes G., Kleijn R., de Koning A., van Oers L., Wegener Sleeswijk A., Suh S., Udo de Haes H. A., de Bruijn H., van Duin R., Huijbregts M. A. J., Lindeijer E., Roorda A. A. H. and Weidema B. P.: *Life cycle assessment; An operational guide to the ISO standards*; Ministry of Housing, Spatial Planning and Environment (VROM) and Centre of Environmental Science (CML), Den Haag and Leiden, Netherlands, 2001, 704p

Table 2.6 Summer SMOG Indicator.

1,1,1-trichloroethane	0.009
1,2,3 trimethylbenzene	1.27
1,2,4 trimethylbenzene	1.28
1,3.5 trimethylbenzene	1.38
1,3 butadiene	0.851
1 butanol	0.62
1 butene	1.08
1 butoxypropanol	0.463
1 butyl acetate	0.269
1 hexene	0.874
1 methoxy 2 propanol	0.355
1 pentene	0.977
1 propanol	0.561
1 propyl benzene	0.636
1 propyl acetate	0.282
1 undecane	0.384
2,2 dimethylbutane	0.241
2,3 dimethylbutane	0.541
2 butadone	0.373
2 butoxyethanol	0.483
2 ethoxyethanol	0.386
2 methoxy ethanol	0.307
2 methyl 1 butene	0.771
2 methyl 2 butene	0.842
2 methylbutan-1-ol	0.489
2 methylbutan-2-ol	0.228
2 methylhexane	0.411
2 methylpentane	0.42
3,5 diethyltoluene	1.3
3,5 dimethylethylbenzene	1.32
3 methyl 1 butene	0.671
3 methylbutane-1-ol	0.433
3 methylbutane-2-ol	0.406
3 methylhexane	0.364
3 methylpentane	0.479
3 pentanol	0.595
Acetaldehyde	0.641
Acetone	0.094
Acetylene	0.085
Acetic acid	0.097
Benzaldehyde	−0.092
Benzene	0.218
Butane	0.352
Butyraldehyde	0.795
Carbon monoxide	0.027
Chloromethane	0.005
Cic-2-butene	1.15
Cis-2-hexene	1.07
Cis-2-pentene	1.12
Cis-dichloroethene	0.447
Cyclohexane	0.29
Cyclohexanol	0.518
Cyclohexanone	0.299

(*Continued*)

Table 2.6 Continued.

Decane	0.384
Diacetone alcohol	0.307
Dichloromethane	0.068
Diethyl ether	0.445
Diethyl ketone	0.414
Diisopropyl ether	0.398
Dimethoxy methane	0.164
Dimethyl carbonate	0.025
Dimethylether	0.189
Dodecane	0.357
Ethane	0.123
Ethanol	0.399
Ethylacetate	0.209
Ethyl-trans-butyl ether	0.244
Ethylbenzene	0.73
Ethene	1
Ethene glycol	0373
Formaldehyde	0.519
Formic acid	0.032
Heptane	0.494
Hexan-2-one	0.572
Hexan-3-one	0.599
Hexane	0.482
Isobutane	0.307
Isobutanol	0.360
Isobutene	0.627
Isobutyraldehyde	0.514
Isopentane	0.405
Isoprene	1.09
Isopropanol	0.188
Isopropyl acetate	0.211
Isopropyl benzene	0.5
Methaethyltoluene	1.02
Metaxylene	1.11
Methane	0.006
Methanol	0.14
Methyl acetate	0.059
Methyl formate	0.027
Methyl isobutyl ketone	0.49
Methyl propyl ketone	0.548
Methyl tert-butyl ether	0.175
Methyl tert-butylketone	0.323
Methyl-isopropylketone	0.364
Neopentane	0.173
NO_2	0.028
NO	−0.427
Octane	0.453
Ortho-ethyltoluene	0.898
Ortho-xylene	1.05
Para-ethyltoluene	0.906
Para-xylene	1.01
Pentanaldehyde	0.765
Pentane	0.395

(Continued)

Table 2.6 Continued.

Propane	0.176
Propanoic acid	0.15
Propionaldehyde	0.798
Propylene	1.12
Propylene glycol	0.457
Sec-butanol	0.447
Sec-butyl acetate	0.275
SO_2	0.048
Styrene	0.142
Tert-butanol	0.106
Tert-butylacetate	0.053
Tetrachloroethene	0.029
Toluene	0.637
Trans-2-butene	1.13
Trans-2-hexene	1.07
Trans-2-pentene	1.12
Trans-dichloroethene	0.392
Trichlorethene	0.325
Trichloromethane	0.023

2.2.1 Ecotoxicity

The characterization factors of substances that contribute to ecotoxicity are listed in table 2.7 (Heijungs, 1992[207]). These factors, which correspond to the opposite of Cm, are as follows depending on the environment concerned:

- ECA (Ecotoxicological Classification factor for Aquatic ecosystems);
- ECT (Ecotoxicological Classification factor for Terrestrial ecosystems).

The table below is only an extract of the CML list, which includes over one hundred substances. Some hydrocarbon values are given, but no values corresponding to pesticides, which do not directly concern the building sector. The indicators are expressed in m^3 of water or in kg of polluted soil (at a maximum tolerable concentration, based on the approach of the US environment protection agency, EPA). They are obtained by adding up the emissions multiplied by the corresponding factors.

A more recent indicator, which integrates the modelling of damage, is presented at §2.3.

2.2.2 Odours

The approach here is similar, with C_m being replaced by a threshold C_s, the odour-detection threshold, defined as the concentration in which 50% of a representative

---

[207]R. Heijungs, *Environmental life cycle assessment of products, Centre of environmental science (CML)*, Leiden, 1992

Table 2.7 Ecotoxicity indicator.

Formula	Substance	ECA	ECT
Metals			
As	arsenic	0.2	3.6
Cd	cadmium	200	13
Cr	chrome	1.0	0.42
Co	cobalt		0.42
Cu	copper	2.0	0.77
Pb	lead	2.0	0.43
Hg	mercury	500	29
Ni	nickel	0.33	1.7
Zn	zinc	0.38	2.6
Hydrocarbons			
C_6H_6	benzene	0.029	
C_6H_5OH	phenol	5.9	5.3
C_6HCl_5O	pentachlorophenol	5.6	5.9
	PCB-28 (polychlorobiphenyl)	16	
	PCB-52	430	
	PCB-101	40	
	PCB-118	360	
	PCB-138	71	
	PCB-153	100	
	PCB-180	130	
	dioxins (TCDD eq.)		1400
$CHCl_3$	chloroform	0.17	
	crude oil	0.05	

sample detects the product. The critical volume is similarly obtained by dividing the emissions by C_s, and the odour indicator by adding up the critical volumes. This indicator is expressed in m^3 of polluted air.

The characterization factors of substances contributing to this phenomenon are listed in table 2.8. These factors, which correspond to C_s, are called OTV (Odour Threshold Value in air) and the following table is only an extract of the CML list, which includes over sixty substances. The indicator is expressed in m^3 of malodorous air (odour equivalent to a concentration of 1 mg/m^3 of ammonia, detected by 50% of a sample of representative people), and is obtained by adding together the emissions (in kg) and dividing it by the corresponding OTV.

2.2.3   Human toxicity

The approach here is more complex, because the effect produced depends on the dose of pollutant received and not its concentration in the environment. A dose is the ratio between the mass of pollutant inhaled or ingested over a certain period and the weight of the individual. A person weighing 70 kg inhales around 20 m^3 of air and ingests around two litres of water per day.

Table 2.8 Odour generation indicator.

Formula	Substance	1/OTV
$C_2H_3Cl_3$	1,1,1-trichloroethane	1.85 10^{+5}
CH_3COCH_3	acetone	13900
CH_3COOH	acetic acid	1.64 10^{+7}
NH_3	ammonia	1.0 10^{+6}
C_6H_5Cl	chlorobenzene	1.0 10^{+6}
CH_2Cl_2	dichloromethane	1560
CS_2	carbon disulphide	5.56 10^{+6}
C_2H_5OH	ethanol	1.56 10^{+6}
CH_3OH	methanol	13700
C_6H_5OH	phenol	2.56 10^{+7}
H_2S	hydrogen sulphide	2.33 10^{+9}
C_2HCl_3	trichlorethene	2.56 10^{+5}

For example, carbon monoxide is toxic starting from long-term exposure at a concentration of 1 g/m^3. The daily dose received per day for a 70 kg individual breathing 20 m^3 of air contaminated with 1 g CO/m^3 is: $20 \times 0.001/70 = 2.8 \ 10^{-4}$ kg CO/kg.

The actual impact in terms of diseases generated depends on the density of the population around the point of emission, and so on the location of the emission. If a pollutant was emitted in the desert and if it degraded rapidly, the population affected would be nil and so would the impact. However, when making a decision, emissions cannot generally be located: when designing a building, no one knows where concrete blocks and bricks will be produced, no one knows whether gas will come from Siberia or Norway in several years' time, nor where electricity will be produced, etc. For this reason, a global average is considered in this simplified indicator.

The Earth's population is P = 6 billion, and the volume of the atmosphere is $V_a = 3 \ 10^{18}$ m^3. The volume V_a considered is restricted to the troposphere, i.e. the equivalent of 6 km high of air to 1 atmosphere (which corresponds to 10 km of actual height). The daily threshold dose D_s considered for each pollutant corresponds, for carcinogenic substances, to 1 additional case of cancer per year for 1000 ha subject to this dose for their whole life. For other diseases, Ds corresponds to the maximum dose at which no effect is observed.

The quantity of human flesh contaminated at dose D_s by the emission E of a substance in the air is: $E \times 20 \times P/V_a/D_s$. A similar reasoning is used for the aquatic environment, taking a depth of 10 m accessible to pollution, i.e. a volume of water $V_e = 3.5 \ 10^{18}$ litres of water on a surface equal to 70% of the surface of the Earth. For emissions in the soil, the weight of the soil P_s that can play a role in toxicity is obtained taking a thickness of 15 cm, 30% of the Earth's surface, and a mass density of 1200 kg/m^3, which results in 2.7 10^{16} kg. Earth is not ingested, and so the exposure to the pollutant is characterized by a parameter p linked to the quantity of pollutant transmitted to a person via water and food. Since the quantity of water and food ingested by a person is supposedly proportional to his or her weight, p corresponds to the average daily weight equivalent of "ingested" soil divided by a person's average weight.

Table 2.9 Human toxicity indicator (extract from Heijungs, 1992[208]).

Formula	Substance	HCA	HCW	HCS
Metals				
As	arsenic	4700	1.4	0.043
Ba	barium	1.7	0.14	0.019
Cd	cadmiun	580	2.9	7
Cr^{3+}	chrome (III)	6.7	0.57	0.018
Cr^{6+}	chrome (VI)	47000	4100	130
Co	cobalt	24	2	0.065
Cu	copper	0.24	0.02	0.0052
Fe	iron	0.042	0.0036	
	iron oxides	0.067	0.0057	
Hg	mercury	120	4.7	0.15
Mn	magnesium	120		
Mo	molybdenum	3.3	0.29	0.7
Ni	nickel	470	0.057	0.014
Pb	lead	160	0.79	0.025
Sn	tin	0.017	0.0014	0.000045
V	vanadium	120		
Zn	zinc	0.033	0.0029	0.007
Non-organic compounds				
NH^{4+}	ammonium	0.02	0.0017	
Br^-	bromide	0.033	0.0029	
CO	carbon monoxidea	0.012		
CN^-	cyanide (free)	0.67	0.057	1.4
	cyanide (complex)	2.6	0.22	5.4
F^-	fluoride	0.48	0.041	
H_2S	hydrogen sulphide	0.78		
NO_3^-	nitrate	0.0091	0.00078	
NO_2^-	nitrate ion	0.26	0.022	
NO_x	nitrogen oxides	0.78		
SO_3^{2-}	sulphites	0.038	0.0033	
SO_2	sulphur dioxide	1.2		
Other				
C_6H_6	benzene	3.9	0.66	
C_6H_5OH	phenol	0.56	0.048	0.62
	chlorobenzene (in general)		0.19	5.7
	chlorophenol (in general, without PCB)	11	0.95	0
$(C_6H_2Cl_2)2O_2$	2,3,7,8 TCDD (dioxin)	3300000	290000	
	chloro-PAH[209] (in general)		67	5.7
	petroleum	1.7	0.00092	
	chloroform	1.2	0.095	3.3

[208] R. Heijungs, *Environmental life cycle assessment of products, Centre of environmental science (CML)*, Leiden, 1992
[209] Polycyclic Aromatic Hydrocarbons

The human toxicity indicator is the sum of the quantities of human flesh contaminated by the different pollutants. It is expressed as:

$$\sum (\text{emissions in air} \times 20 \times P/V_a/D_s) + \sum (\text{emission in water} \times 2 \times P/V_e/D_s)$$
$$+ \sum (\text{emissions in soil} \times p \times 70 \times P/P_s/D_s)$$

The characterization factors of substances that contribute to this phenomenon are listed in table 2.9. These factors are as follows depending on the environment concerned (expressed in kg of flesh contaminated at D_s/kg of substance emitted):

- HCA (Human toxicological Classification factor for Air);
- HCW (Human toxicological Classification factor for Water);
- HCS (Human toxicological Classification factor for Soil).

The table below is only an extract of the CML list, which includes over 100 substances. As for ecotoxicity, some values are given for hydrocarbons, but no values are given that correspond to pesticides. The indicator is expressed in kg of human flesh contaminated at a tolerable maximum dose. It is found by adding the quantities emitted in the air (or water and soil), and multiplying by the HCA (or HCW or HCS).

Once again, the list of substances considered here is not exhaustive. Some products, such as glycol ethers, do not feature, despite their noxiousness: these products dissolve pigments in some paints and cannot be sold to the general public but can be used by professionals. This example illustrates the limitations of indicators, which should evolve to include the numerous new substances available on the market.

A more recent indicator that integrates modelling of damage is described in §2.3.

2.2.4   Indicators of the AFNOR standard NF P01-010

The AFNOR standard NF P01-010 on the environmental quality of building products[210] includes an annex on how to calculate "environmental impacts" and defines a number of indicators, most of which are similar to the indicators presented here (climate change, acidification, destruction of the ozone layer, formation of photochemical ozone, consumption of resources, waste). Two indicators are specific to this standard:

- An indicator of air pollution, based on the critical volume method taking the threshold concentration for each substance as authorized by the decree of 2 February 1998 (a new version was compiled in 2007);
- An indicator of water pollution based on the same method, with a threshold corresponding to the same decree but for concentration in water.

These two indicators are expressed respectively in m^3 of air and m^3 of polluted water. The threshold values correspond to a French decree, so these indicators would

---

[210]AFNOR standard NF P01-010, Qualité environnementale des produits de construction – Déclaration environnementale et sanitaire des produits de construction, December 2004

be more difficult to apply in other countries than those based on less specific thresholds. Dioxins, for example, do not feature among the substances considered in the AFNOR air pollution indicator (they may be taken into account as an unspecified volatile organic compound), whereas in the CML's human toxicity indicator, they are considered (in 2,3,7,8 TCDD equivalent) to be 275 million times more toxic than carbon monoxide. However, the French law sets a concentration limit at 0.1 ng/m^3 (transposition of the European directive 76/CE of 4 December 2000), which is 550 million times more restrictive than for carbon monoxide (restricted to 0.055 g/m^3 in the work place). The VOC threshold applied in the AFNOR standard is 0.15 g/m^3, which is 2.7 times less restricting than for CO. Dioxins (in 2,3,7,8 TCDD equivalent) are considered as VOCs in the AFNOR standard, and so evaluated as 2.7 times more toxic in the CML indicator, and 550 million times more in the DALY indicator (cf. below). The same goes for phtalates, which are 2000 times more toxic than CO (in phtalate-dioctyl equivalent) in the DALY indicator.

The water pollution indicator groups aspects linked to eco-toxicity (e.g. heavy metals) with others linked to eutrophication. The advantage is that this reduces the number of indicators, but the disadvantage is that the role of different actors can no longer be dissociated: e.g. households can act on eutrophication by choosing certain washing powders and dry toilets, and some industrials can act on heavy metal emissions.

2.3   DAMAGE-ORIENTED INDICATORS

The preceding indicators are linked to environmental issues (e.g. climate change, acid rain). Damage-oriented indicators aim to integrate more downstream effects, such as health problems linked to climate change, ionizing radiation, changes in the ozone layer, or damage to the quality of ecosystems. They are sometimes termed "end-point" indicators as opposed to the previously mentioned "mid-point" indicators: the aim is to assess final, rather than intermediate, impacts. They require more advanced models[211], and research continues in this domain. One indicator, expressed in years of life lost, is used to evaluate the impact in terms of public health, another concerns ecotoxicity[212].

2.3.1   Impacts on health (years of life lost)

Health impacts combine diseases and deaths into a single indicator called DALY (disability adjusted life loss). This involves establishing an equivalent between a year of illness and a year of life lost depending on the seriousness of the illness. The table below

[211] For example, EUSES, the European Union System for the Evaluation of Substances, developed by the National Institute of Public Health and the Environment (RIVM) in the Netherlands (EUSES, 2008) cf. https://ec.europa.eu/jrc/en/scientific-tool/european-union-system-evaluation-substances, IMPACT 2002 developed at the Ecole Polytechnique Fédérale de Lausanne (Jolliet, 2003) and Impact Pathways developed at the Ecole des Mines de Paris (Freidrich, Rabl and Spadaro, 2001)

[212] M.J. Goedkoop and R. Spriemsma, *The Eco-Indicator 99, A damage oriented method for life cycle impact assessment, methodology report, methodology annex, manual for designers*, April 2000

Table 2.10 DALY equivalents for different disabilities.

Disability	DALY equivalent (D)
Vitiligo (skin condition), weight loss	0 to 0.02
Diarrhoea, sore throat, anaemia	0.02 to 0.12
Fracture, infertility, arthritis, angina	0.12 to 0.24
Amputation, deafness	0.24 to 0.36
Recto-vaginal fistula, moderate mental retardation, trisomy	0.36 to 0.5
Serious depression, blindness, paraplegia	0.5 to 0.7
Psychosis, dementia, severe migraine, quadriplegia	0.7 to 1

shows the hypotheses considered by Murray and Lopez[213] for such an equivalent, marked D (0 = perfect health, 1 = death).

The same authors suggest taking into account the age of the ill person to evaluate the number of years lived disabled (YLD). To do this, they introduce three parameters: an age modulation factor, K, a discount rate, r, and an age weighting parameter, β. The expression for calculating YLD is thus:

$$YLD = D \left\{ \frac{KCe^{ra}}{(r+\beta)^2} [e^{-(r+\beta)(L+a)}[-(r+\beta)(L+a) - 1] - e^{-(r+\beta)a}[-(r+\beta)a - 1]] \right.$$
$$\left. + \frac{1-K}{r} \left(1 - e^{-rL}\right) \right\}$$

where C is a fixed adjuster, equal to 0.1658;
a is the age at which the disability started;
and L is the length of the disability.

The graph below shows how the YLD indicator varies depending on age, for a one-year disability equivalent to 0.5 DALY. Two cases are compared:

-   the modulation depending on age (where K = 1, r = 3% and β = 4%),
-   a constant value whatever the age.

The discount rate has the effect of reducing long-term impacts, which gives the following curve: if these rates are not nil, then YLD is not proportional to the duration of the disability.

The authors apply the same principle to the years of life lost, YYL, taking D = 1 and a "disability" duration equal to life expectancy minus the age when death occurred:

$$YLL = \frac{KCe^a}{(r+\beta)^2} [e^{-(r+\beta)(L+a)}[-(r+\beta)(L+a) - 1] - e^{-(r+\beta)a}[-(r+\beta)a - 1]]$$
$$+ \frac{1-K}{r}(1 - e^{-rL})$$

[213] Murray C. and Lopez, A.: The Global Burden of Disease, WHO, World Bank and Harvard School of Public Health, Boston, 1996, 990p

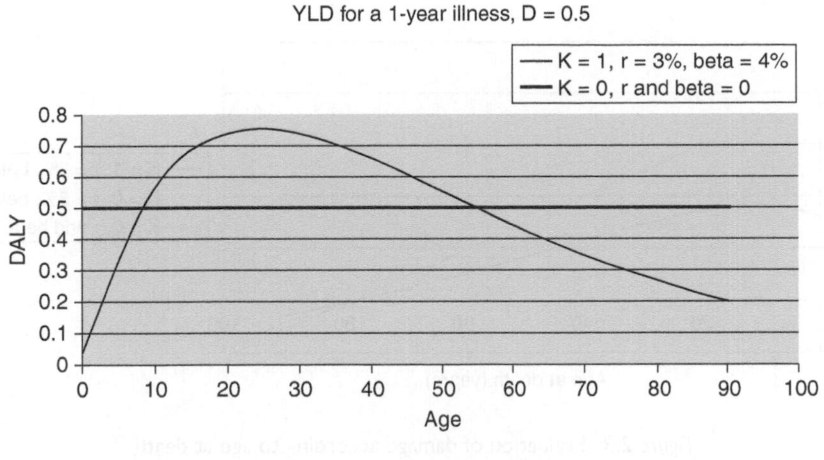

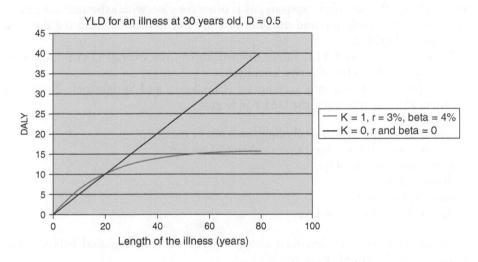

YLD for a 1-year illness, D = 0.5

— K = 1, r = 3%, beta = 4%
— K = 0, r and beta = 0

Figure 2.1 Modulation of damage evaluation according to age.

YLD for an illness at 30 years old, D = 0.5

— K = 1, r = 3%, beta = 4%
— K = 0, r and beta = 0

Figure 2.2 Effect of a discount on evaluating damage.

a being the age at the time of death;
and L the life expectancy at age *a* (i.e. at birth 82.5 years for women and 80 years for men in this study).

The modulation according to age reduces the number of years counted according to the discount rate considered, which gives the following graph (taking life expectancy at birth to be 80 years).

The use of modulation according to age raises ethical issues: should we really consider that a year of a child's life or an old person's life is worth less than a year of the life of an apparently more productive adult? Similarly, the use of a discount rate reduces the amount of years of life lost in a more distant future (e.g. if the illness only

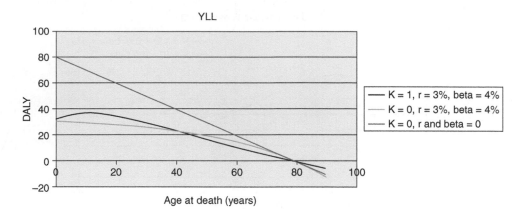

Figure 2.3 Evaluation of damage according to age at death.

emerges 10 or 20 years after exposure, as is often the case with asbestos and cancer). As a result, the modulation and discount rate are not implemented in the databases developed for EQUER.

Taking the YLD and YLL values defined above, the overall DALY indicator is obtained by simply adding the two quantities.

Six aspects are taken into account by (Goedkoop and Spriemsma, 2000)[214] to evaluate health impacts using the DALY indicator:

– carcinogenic substances (including some heavy metals),
– summer smog (formation of photochemical ozone),
– winter smog (dust, SO_2),
– climate change,
– ionizing radiation,
– alterations in the ozone layer.

Unlike the indicators described above based on a simple global balance sheet, damage-oriented methods must model:

– the fate of substances in the environment

    – emissions in ecological compartments (air, freshwater, seawater, sediments, natural, agricultural and industrial soil)
    – (bio)-degradation (hydrolysis, photolysis in water, photochemical reactions in the atmosphere, biodegradation in water, sediments and soil)
    – transport: diffusion (air, water), absorption/volatilization (air/water or soil), adsorption/desorption (air/soil), dry deposition (air/soil) and wet deposition

---

[214]M.J. Goedkoop and R. Spriemsma, *The Eco-Indicator 99, A damage oriented method for life cycle impact assessment, methodology report, methodology annex, manual for designers*, April 2000

(rain/soil), sedimentation/resuspension (sediments/water), runoff, erosion (soil/water)
- concentration in ecological compartments,
- transferral into the air, water (purification), food (bioconcentration in plants, bioaccumulation in animals, biomagnification via the food chain)
- the doses received by people
  - breathing (23, or 21 and 15 m^3/day for men, women and children), ingestion of water (2, or 1.4 l/day) and food (2.5 kg/day), transmission through the skin,
  - doses received by the population concerned (local, regional, global/men, women, children/workers, non-professionals).
- the diseases generated
  - dose-response factor used to evaluate the risk of disease according to the dose, with a threshold below which no effect is observed (e.g. in respiratory disease) or with no threshold (e.g. cancer),
  - years of life lost or with disability.

The models developed for these evaluations are still patchy (e.g. interactions between substances are not integrated) and they require numerous data on substances (physico-chemical properties linked to solubility, degradation, etc., dose-response factors, etc.). Damages depend on the location of emissions (climate aspects like wind direction and speed over time, population density and duration of exposure, background concentration of pollutants, etc.). However, an order of magnitude corresponding to an average constitutes a first step, and current research on this subject will hopefully make progress.

a)   carcinogenic substances (including some heavy metals),

The substances considered correspond to the following classes (established by the International Agency for Research on Cancer): 1 (substances carcinogenic to humans), 2A (substances probably carcinogenic), 2B (substances possibly carcinogenic) and 3 (substances not classifiable). In other words, only category 4 (non-carcinogenic substances) is not taken into account.

The model used for the dispersion of pollutants and their fate in the different ecological compartments is the European Uniform System for the Evaluation of Substances (EUSES, 2008)[215]. This model makes it possible to estimate, for a given emission, a substance's concentration in the air, water and food, and then deduce the doses received by the population concerned. The population concerned depends on the pollutant's lifespan in the atmosphere, and the hypotheses are as follows (interpolations have been made between the different cases):
- the population density is that of Western Europe (160 people/km^2) if the time spent in the atmosphere is one day, taking dispersion as throughout Europe,
- the density considered is the global average if the time spent is one year,
- the density is at its maximum (300 people/km^2) for longer periods.

---

[215] https://ec.europa.eu/jrc/en/scientific-tool/european-union-system-evaluation-substances

The dose-response factors are evaluated from risk factors, corresponding to the risk generated from exposure to a concentration of 1 $\mu g/m^3$ for 70 years. The risk is taken to be proportional to the dose. Only the most frequent type of cancer is considered, since dose-response factors do not break down into different types. The YLD and YLL values are obtained from statistics on different types of cancer.

b)   summer smog and winter smog

The substances involved in summer smog (VOC, NOx, Ozone, etc.) and winter smog (dust, SO_2) are modelled in the form of a relationship between emissions and concentration, taking non-linear mechanisms into account for ozone. Doses are then estimated for the populations concerned, followed by the dose-response functions (e.g. linking the number of hospital stays to doses). The equivalent number of years of life lost is then deducted using DALY equivalents.

c)   climate change

Temperature variations are a cause of death: 15,000 additional deaths were observed in France during the summer 2003 heat wave, mostly due to respiratory and cardio-vascular problems. Climate change can also provoke the spread of infectious diseases like malaria and dengue fever. The FUND model also evaluates loss of territory linked to rising water and resulting migrations, reduction in agricultural yields, the effects of cyclones, storms and floods in 9 regions in the world. An evaluation of these effects has been carried out for the period from 2000 to 2200 using GIEC's IS92a scenario, then adding 1 Mt of CO_2, CH_4 and N_2O so as to evaluate the marginal impact in DALYs of these reference emissions: Dref. Some DALY values can be negative, e.g. the number of cardio-vascular accidents can drop in cold weather, but positive and negative values are only aggregated in a single region to avoid undue compensations.

The results of this model are extrapolated from other greenhouse gas effects, taking the effects to be the same as methane if the lifespan of the gas is below 20 years, the same as CO_2 for a lifespan of between 20 and 110 years, and as nitrogen protoxide for gases with a lifespan above 110 years. The Di damage linked to the emission of one kg of substance i is thus expressed by:

$$D_i = D_{ref} \cdot GWP_i/GWP_{ref}$$

where GWP_i and PWP_{ref} are respectively the potential for global warming of substance i and of the reference substance.

d)   ionizing radiation

Taking the emission of radiation at a given place, expressed in Becquerel, the model estimates the contamination of the environment, taking into account the transport, dispersion and deposition of contaminated matter. An exposure model evaluates a dose in Sievert according to the irradiation conditions, the consumption of contaminated water and food, etc. Dose-response functions are then used to estimate the number of cancers and hereditary diseases, which are transformed into the DALY scale.

e)   alteration of the ozone layer

The alteration of the ozone layer is evaluated from the annual reduction in ozone concentration, expressed in ppb (parts per billion) per year. This reduction

Table 2.11   DALY indicator (disability adjusted years of life lost).

Emissions in the air	
Acetaldehyde	$1.56 \ 10^{-6}$
Acetone	$2.4 \ 10^{-7}$
Acetic acid	$2.13 \ 10^{-7}$
Aldehydes	$1.4 \ 10^{-6}$
Ammonia	$8.5 \ 10^{-5}$
Arsenic	$2.46 \ 10^{-2}$
Benzene	$2.97 \ 10^{-6}$
Benzo-a-pyrene	$3.98 \ 10^{-3}$
Butadiene	$1.77 \ 10^{-5}$
Butane	$1.39 \ 10^{-6}$
Butene	$2.47 \ 10^{-6}$
Cadmium	$1.35 \ 10^{-1}$
Carbon dioxide (including greenhouse effect)	$2.1 \ 10^{-7}$
Carbon monoxide	$3.22 \ 10^{-7}$
Cesium 134	$1.2 \ 10^{-8}$
CFC 10	$1.84 \ 10^{-3}$
CFC 11	$1.27 \ 10^{-3}$
CFC 12	$2.26 \ 10^{-3}$
CFC 13	$4.17 \ 10^{-3}$
CFC 113	$1.58 \ 10^{-3}$
CFC 114	$3.08 \ 10^{-3}$
CFC 115	$2.71 \ 10^{-3}$
CFC 143a	$7.8 \ 10^{-4}$
Chloroform	$2.72 \ 10^{-5}$
Chrome VI	$5.84 \ 10^{-3}$
Cobalt 60	$1.6 \ 10^{-8}$
Dioxin 2-3-7-8 TCDD	$1.79 \ 10^{2}$
1,2 dichloro-ethane	$3.02 \ 10^{-5}$
ethane	$2.64 \ 10^{-7}$
ethanol	$8.34 \ 10^{-7}$
FC 14	$1.4 \ 10^{-3}$
formaldehyde	$2.1 \ 10^{-6}$
halon 1001	$6.74 \ 10^{-4}$
halon 1211	$5.64 \ 10^{-3}$
halon 1301	$5.5 \ 10^{-3}$
HCFC 21	$8.64 \ 10^{-5}$
HCFC 22	$3.22 \ 10^{-4}$
HCFC 31	$3.5 \ 10^{-7}$
HCFC 123	$2.13 \ 10^{-5}$
HCFC 124	$1.17 \ 10^{-4}$
HCFC 140	$8.3 \ 10^{-5}$
HCFC 141b	$1.57 \ 10^{-4}$
HCFC 142	$3.93 \ 10^{-4}$
HFC 23	$2.6 \ 10^{-3}$
HFC 32	$1.4 \ 10^{-4}$
HFC 116	$2 \ 10^{-3}$
HFC 125	$5.8 \ 10^{-4}$
HFC 134a	$2.7 \ 10^{-4}$

(Continued)

Table 2.11 Continued.

Emissions in the air	
HFC 152a	$2.9 \ 10^{-5}$
krypton 85	$1.4 \ 10^{-13}$
lead 210	$1.5 \ 10^{-9}$
methane	$4.41 \ 10^{-6}$
methanol	$2.81 \ 10^{-7}$
nickel	$4.29 \ 10^{-5}$
N_2O	$6.9 \ 10^{-5}$
NOx	$8.87 \ 10^{-5}$
PAH (polycyclic aromatic hydrocarbons)	$1.72 \ 10^{-4}$
paraffin	$1.28 \ 10^{-6}$
particles $< 2.5 \ \mu m$	$7.1 \ 10^{-4}$
particles from 2.5 to $10 \ \mu m$	$3.75 \ 10^{-4}$
particles $> 10 \ \mu m$	0
pentachlorophenol	$7.21 \ 10^{-3}$
plutonium 238	$6.7 \ 10^{-8}$
plutonium alpha	$8.3 \ 10^{-8}$
polonium 210	$1.5 \ 10^{-9}$
polychlorobiphenyls (PCB)	$1.97 \ 10^{-3}$
propane	$1.1 \ 10^{-6}$
radium 226	$9.1 \ 10^{-10}$
radon 222	$2.4 \ 10^{-11}$
sodium dichromate	$2.32 \ 10^{-3}$
sulphates and SO_2	$5.46 \ 10^{-5}$
SF_6	$5.3 \ 10^{-3}$
trichlorethene	$6.98 \ 10^{-7}$
toluene	$1.36 \ 10^{-6}$
uranium 234	$9.7 \ 10^{-8}$
uranium 235	$2.1 \ 10^{-8}$
uranium 238	$8.2 \ 10^{-9}$
xenon	$1.4 \ 10^{-13}$
xylene	$2.21 \ 10^{-6}$
Emissions in water	
Antimony 124	$8.2 \ 10^{-10}$
Arsenic	$6.57 \ 10^{-2}$
Benzene	$4.12 \ 10^{-6}$
Cadmium	$7.12 \ 10^{-2}$
Carbone 14	$1.2 \ 10^{-9}$
Chloroform	$2.6 \ 10^{-5}$
Chrome VI	$8.26 \ 10^{-10}$
Cobalt 60	$4.4 \ 10^{-8}$
Formaldehyde	$4.97 \ 10^{-6}$
Manganese 54	$3.1 \ 10^{-10}$
Monochloro-methane (R40)	$2.11 \ 10^{-5}$
Nickel	$6.91 \ 10^{-11}$
PAH (polycyclic aromatic hydrocarbons)	$2.6 \ 10^{-3}$
Phtalate-dioctyl	$6.64 \ 10^{-4}$

(Continued)

Table 2.11 Continued.

Emissions in water	
plutonium alpha	$7.4 \cdot 10^{-9}$
radium 226	$1.3 \cdot 10^{-10}$
Silver	$5.1 \cdot 10^{-10}$
Styrene	$1.22 \cdot 10^{-6}$
Uranium 234	$2.4 \cdot 10^{-9}$
Uranium 235 and 238	$2.3 \cdot 10^{-9}$
Emissions in soil	
Arsenic	0.0132 to 0.25
Cadmium	$3.98 \cdot 10^{-3}$ to 2.17
Chrome VI	$3.68 \cdot 10^{-7}$
Lindane	$8.64 \cdot 10^{-3}$
Nickel	$4.21 \cdot 10^{-9}$

is modelled as proportional to the difference between actual chlorine concentration and concentration resulting in no damage, evaluated at 1.9 ppb. The factor of proportionality depends on the latitude, its average value being 2% at the most populated latitudes, i.e. between 30° South and 55° North.

The reference substance is CFC11, $CFCl_3$. An increase in chlorine concentration is taken to be proportional to the mass of CFC11 emitted, to the number of chlorine atoms (3 in the case of CFC11), and to a factor expressing the alteration of the ozone layer per kg of pollutant emitted. Equivalent coefficients, ODP (Ozone Depletion Potential), have been defined for substances other than CFC11 (Hauschild and Wenzel, 1998).

The increase in ultra-violet radiation is deduced from the alteration of the ozone layer using a law of proportionality, and dose-response functions are then used to deduce the health risk, so as to evaluate a DALY indicator. Three-quarters of this value correspond to years with a sight disability (i.e. cataracts), the other quarter is mainly linked to the risk of skin cancer.

The DALY values corresponding to the emission of 1 kg in the air, water or soil are given below for different substances (Goedkoop and Spriemsma, 2000), but this type of indicator remains highly uncertain.

Note that in the definition of this indicator, dioxin 2-3-7-8 TCDD is around 100 million times more toxic than formaldehyde (and over 500 million times compared to CO). It is therefore not very precise to class this substance under non-specified volatile organic compounds, as recommended by the AFNOR standard NF P01 010 (used to draw up FDES declarations).

The DALY indicator is incomplete, for instance it does not take into account some toxic effects linked to heavy metals: e.g. mercury emissions, despite the fact that the Minamata catastrophe illustrated their toxicity. Research is therefore necessary to improve the relevance of toxicity indicators.

As an illustration, the table below shows the values given by this indicator for some materials and procedures, obtained from the Ecoinvert database (version 2003).

Table 2.12 DALY indicator for some materials and procedures (calculated using the econinvent database, version 2003).

Materials (per kg)	DALY
manufacture 1 kg concrete	7.00E–07
manufacture 1 kg steel reinforcement	6.72E–06
manufacture 1 kg new high alloy steel	2.12E–05
manufacture 1 kg glass	4.02E–06
manufacture 1 kg water-based paint	4.89E–06
manufacture 1 kg solvent-based paint	5.61E–06
manufacture 1 kg polyurethane	7.77E–06
manufacture 1 kg PVC	4.15E–06
manufacture 1 kg polystyrene	9.56E–06
Energy	
heating wood (MWh)	5.83E–05
heating gas (MWh)	8.72E–06
heating fuel oil (MWh)	1.60E–05
electricity from nuclear (MWh)	4.53E–06
electricity from gas (MWh)	2.36E–05
electricity from coal (MWh)	8.67E–05
Other	
waste water treatment (1 person year)	3.23E–04
transport by car (1 person km)	3.29E–07
transport by lorry (1 tonne km)	6.75E–07
incineration of 1 kg cellulose	7.23E–07
incineration of 1 kg paint (special waste)	1.55E–06
incineration of 1 kg PVC	2.26E–06
incineration of 1 kg wood	1.42E–07

Apart from the limits already indicated, risks are not taken into account in the DALY indicator, which corresponds to toxicity under normal use. This indicator should therefore be considered as an incentive to different industries to improve their performance rather than an absolute evaluation. For example, new wood boilers include anti-pollution devices, and their performance is better than that shown in the above table, which corresponds to an average that includes old equipment. The above values are therefore likely to evolve considerably, partly due to research on refining the modelling of this very vast set of phenomena linking damage to emissions, and partly thanks to improved products and processes.

2.3.2 Impacts on ecosystems

The EUSES model described above takes emissions from pollutants in the air, water and industrial and agricultural soil and evaluates their transferral into other ecological compartments, such as surface waters and interstitial water in natural, agricultural and industrial soils. A concentration can thus be estimated in these compartments and linked to the NOEC (no observed effect concentration), which is obtained by toxicity tests on different species. A percentage of potentially effected species (PAF: potentially affected fraction) is then deduced for each pollutant using dose-response functions. A background concentration corresponding to a European average is used to evaluate the marginal impact caused by an additional emission. Interactions between pollutants are only taken into account for water repellent and inert substances. The indicator is ultimately expressed as a percentage of potentially affected species, over a certain surface area (in the EUSES model, each ecological compartment has a certain surface) and over a certain length of time. The corresponding unit is PAF $* m^2 *$ year. The value in PAF can be transformed into PDF, Potentially Disappeared Fraction, with the hypothesis: 1 PDF $= 10$ PAF. This transformation can integrate other causes of disappearing species, indicated below.

–   damage caused by acidification and eutrophication

The SMART dynamic model was developed in the Netherlands to monitor the evolution of pollutants like NO_x, SO_x and NH_3, integrating the nitrogen cycle, biochemical processes and a simplified hydrological model. The model evaluates changes in the soil's acidity and nitrogen concentration on an annual basis, responding to a marginal emission of pollutants on 1 km^2 grid boxes.

The MOVE model, also developed in the Netherlands, contains dose-response functions for 900 plant species. Plants in danger are counted on each grid box over time. A PDF*m^2*year value is deduced by aggregation on the Netherlands territory, depending on the persistence of the different pollutants. The results are extrapolated on a European scale taking 60% of the territory to be "natural".

–   damage caused by land use and transformation

Biodiversity, expressed by the number of species S on a given territory, is linked to the surface A of this territory using a standard expression:

$$S = a * A^b$$

where a is a "species richness factor", generally between 20 and 2000,
b is a "species accumulation factor", generally between 0.2 and 0.5.

The values of a and b are obtained for different types of soil using field studies (Kollner, 1999[216]). The potentially disappeared fraction (PDF) is defined from the difference between the number of species before and after land use or conversion, divided by the number of species before this transformation. "Natural" soil is taken to be the initial condition in cases of land use without conversion. The number of species thus diminishes in a portion of territory that remains natural, but whose surface area diminishes. Two effects are therefore cumulated in the indicator.

The PDF value is multiplied by the land surfaces concerned, and by the duration of the use, or of the restoration of the initial state in the case of a conversion. This restoration duration is taken to be 5 years in the case of conversion of agricultural/urban land, and 30 years in the case of conversion of natural/urban or agricultural land. Given that the field studies are carried out on the fringes of territories, the number of disappeared species is multiplied by 2 as a precaution, to take into account the greater biodiversity at the centre of territories.

---

[216]Köllner, T.: Species-pool Effect Potentials (SPEP) as a yardstick to evaluate land-use impacts on biodiversity, Journal of Cleaner Production, August 1999

Table 2.13   Indicator PDF $\cdot$ m^2 $\cdot$ year (potentially disappeared fraction).

Emissions in the air	
Ammonia	15.6
Arsenic	592
Benzene	0.00275
Benzo-a-pyrene	142
Cadmium	9650
chrome	4130
copper	1460
Dioxine 2-3-7-8 TCDD	132 000
lead	2540
mercury	829
nickel	7100
nitrates and NOx	5.71
PAH (polycyclic aromatic hydrocarbons)	7.8 10^{-4}
pentachlorophenol	13.3
sodium dichromate	1640
sulphates and SO$_2$	1.04
toluene	2.4 10^{-4}
zinc	2890
Emissions in water	
Arsenic	11.4
Benzene	0.048
Cadmium	480
Chrome	68.7
copper	147
dichromate	33.1
lead	7.39
mercury	197
Nickel	143
PAH (polycyclic aromatic hydrocarbons)	0.0021
Phtalate-dibutyl	1.62
toluene	0.173
zinc	16.3
Emissions in soil	
Arsenic	2.16 or 610*
Atrazine	0.149
cadmium	30.1 or 9940*
Chrome	54.8 or 4240*
lead	2.82 or 12.9
mercury	272 or 1680
Nickel	20.8 or 7320
Zinc	2.32 or 2980
Land usage	
Use of one m^2 for one year	0.1 to 1.15
Transformation of one m^2,	
Arable –> other usage	−34.4
Other usage –> arable	34.4
Landfill –> other usage	−25 to −28

*the lowest value corresponds to an agricultural emission (they have lower concentrations)

2.4 OTHER INDICATORS

2.4.1 Known abiotic resources

The resources of a basin or geographical zone, also called "in-place volumes", are the total quantities of matter present in fields that have been or remain to be discovered in the region considered, without any technical or economic consideration. The reserves are the quantities that operators hope to extract and exploit profitably in the near future. The transition from resource to reserve is characterized by the rate of recuperation. Reserves can be subdivided into four categories:

- proven reserves (exploitation may be underway or not, probability is higher than 85–95% depending on the organization);
- probable reserves (probability estimated at 50%, exploitable on technical and economic conditions in the near future);
- possible reserves (presence estimated at a probability of 5 to 10%, exploitable in the undetermined future);
- expected reserves are defined by weighting the different preceding types, e.g. proven reserves +2/3 (or 1/2) probable reserves +1/3 (or 1/4) possible reserves.

To give an idea of magnitude, proven oil reserves are around 1000 billion barrels[217] (138 billion tonnes, i.e. less than 40 years of current consumption) and expected reserves are twice that size.

Standard indicators only consider those resources that may become insufficient in the next 100 years. Table 2.13 presents the values of their reserves as considered by CML.

The reserve depletion indicator is obtained by adding the quantities of raw materials used to manufacture the product studied and dividing by the reserves. It is therefore dimensionless.

Another indicator is defined by dividing each quantity of matter by the corresponding number of years of reserves. A third possibility is to divide both by the reserves and by the number of years of reserves, so as to attract attention to relatively rare raw materials that are depleting fast.

The indicators defined above result in extremely low values, since the quantity used for a given product is low in relation to world reserves. An indicator in antimony equivalent has thus been defined to produce more reasonable values. The characterization factors corresponding to "ultimate" reserves (i.e. over and above economically exploitable reserves) and integrating the speed of extraction are shown in the table below in antimony equivalent per kg (per m^3 for natural gas). The characterization factor (Abiotic Depletion Potential, in antimony equivalent per kg) of resource i is calculated using the expression below.

$$ADP_i = \frac{DR_i}{(R_i)^2} \cdot \frac{(R_{ref})^2}{DR_{ref}}$$

---

[217]Denis Barbusiaux and Jean Coiffard, *Les combustibles fossiles : quels usages, quelles réserves*, La jaune et la rouge, 2000

Table 2.14 Reserve depletion indicator.

Formula	Substance	Reserves[218]	Unit	Years[219]
Energy resources				
–	crude oil	123,559	Megatonne	75
–	natural gas	109,326	10^9 m^3	66
U	uranium[220]	1,676,820	tonne	48
U	uranium[221]	13,410,000		383
Metals				
Cd	cadmium	0.535	Megatonne	60
Cu	copper	350	Megatonne	55
Hg	mercury	0.0057	Megatonne	92
Ni	nickel	54	Megatonne	120
Pb	lead	75	Megatonne	45
Sn	tin	4.26	Megatonne	56
Zn	zinc	147	Megatonne	56

R_i = ultimate reserve of resource i, [kg];

DR_i = extraction rate of resource i, [kg $*$ year^{-1}];

R_{ref} = ultimate reserve of antimony, $4.63 * 10^{15}$ kg;

DR_{ref} = extraction rate of antimony, $6.06 * 10^7$ kg/an.

Table 2.15 Resource depletion indicator (antimony equivalent), excerpt from (Guinée ed., 2001)[222].

Aluminium	10^{-8}
Antimony	1
Argon	4.71 10^{-7}
Arsenic	0.00917
Bismuth	0.0731

(Continued)

In the above indicator, the ultimate reserves considered for uranium are 62,500 billion tonnes, which does not correspond to the size in other bibliographic references

[218] R. Heijungs, *Environmental life cycle assessment of products*, *Centre of environmental science (CML)*, Leiden, 1992

[219] U.S. Bureau of Mines, 1998

[220] R. Heijungs, *Environmental life cycle assessment of products*, *Centre of environmental science (CML)*, Leiden, 1992

[221] U.S. Bureau of Mines, 1998

[222] Guinée J. B., (final editor), Gorrée M., Heijungs R., Huppes G., Kleijn R., de Koning A., van Oers L., Wegener Sleeswijk A., Suh S., Udo de Haes H. A., de Bruijn H., van Duin R., Huijbregts M. A. J., Lindeijer E., Roorda A. A. H. and Weidema B. P.: *Life cycle assessment; An operational guide to the ISO standards*; Ministry of Housing, Spatial Planning and Environment (VROM) and Centre of Environmental Science (CML), Den Haag and Leiden, Netherlands, 2001, 704p

Table 2.15 Continued.

Bore	0.00467
Brome	0.00667
Cadmium	0.33
Calcium	$7.08 \ 10^{-10}$
Cesium	$1.91 \ 10^{-5}$
Chlorine	$4.86 \ 10^{-8}$
Chrome	0.000858
Coal	0.0134
Cobalt	$2.62 \ 10^{-5}$
Copper	0.00194
Crude oil	0.0201
Fluoride	$2.96 \ 10^{-6}$
Gold	89.5
Helium	148
Iodine	0.0427
Iron	$8.43 \ 10^{-8}$
Krypton	20.9
Lead	0.0135
Lignite	0.00671
Lithium	$9.23 \ 10^{-6}$
Magnesium	$3.73 \ 10^{-9}$
Manganese	$1.38 \ 10^{-5}$
Mercury	0.495
Molybdene	0.0317
Natural gas (m3)	0.0187
Neon	0.325
Nickel	0.000108
Palladium	0.323
Phosphorus	$8.44 \ 10^{-5}$
Platinum	1.29
Polonium	$4.79 \ 10^{14}$
Potassium	$3.13 \ 10^{-8}$
Radium	$2.36 \ 10^{7}$
Selenium	0.475
Silicon	$2.99 \ 10^{-11}$
Silver	1.84
Sodium	$8.24 \ 10^{-11}$
Sulphur	0.000358
Tantalum	$6.77 \ 10^{-5}$
Tellurium	52.8
Tin	0.033
Titanium	$4.4 \ 10^{-8}$
Tungsten	0.0117
Uranium	0.00287
Vanadium	$1.16 \ 10^{-6}$
Xenon	17 500
Zinc	0.000992
Zirconium	$1.86 \ 10^{-5}$

(i.e. from 2 to 13 million tonnes for expected reserves). The relevance of the notion of ultimate reserves is debateable: an indicator based on probable reserves would probably be more appropriate.

Other indicators, expressed in MJ, are based on an exergetic equivalent or on the energy needed to extract resources (Ecoindicators 99). However, reducing the problem of resource depletion to the energy dimension alone is once more debateable. In addition, uranium is not included in the MJ indicator of the Ecoindicator 99 method. Yet evaluating exploitable uranium reserves is actually linked to the dilution of this resource, and to the fact that below a certain concentration, more energy is needed to extract it than the potential recuperation by the nuclear electricity industry. Once again, work needs to be done to improve indicators.

A similar indicator has been envisaged for biotic reserves, but its precise definition has not yet been revealed.

2.4.2 Primary energy

Choosing a primary energy indicator

Taking the yields of the different energy chains, more starting energy is required to produce one kWh of electricity consumed at the meter ("final" kWh) than to produce one final kWh of heat (e.g. gas or oil). Therefore, adding together the consumptions of final energy is not a good way to evaluate the depletion of energy resources. With the objective of taking into account different types of energy (electricity, heat) on a homogeneous basis, a primary energy indicator has thus been defined.

The principle of a life cycle assessment involves going back to the extraction phase of fuel (crude oil, uranium, gas, coal, etc.) or other resources (hydroelectricity). For example, to produce one kWh of electricity requires extracting a certain quantity of uranium, oil and gas, and makes use of hydraulic energy. These different quantities are then translated into primary energy (MJ) and can be added together on a homogenous basis.

Choosing a higher calorific value

The distinction between LCV and HCV (lower and higher calorific value) only makes sense for actual fuel (e.g. gas, fuel oil and coal). It is much harder to define for uranium, since it involves assessing the recoverable energy using realistic technical principles: although it is relatively easy to reply to the question: "Is it realistic to want to recover energy from smoke condensation?" (choosing the HCV implicitly implies that the answer is yes), it is harder to know what share of the energy contained in natural uranium can be potentially used. The convention considered in the Oekoinventare[223] database (developed by the Ecole Polytechnique Fédérale de Zurich) involves adding up the energy contained in enriched and depleted uranium. Natural uranium is supposed to contain 0.7% of uranium U_{235}, and enriched uranium to contain 3.5%. One kg of uranium U_{235} can produce $128\,TJ_e$ (electric terajoules, 1 TJ is

---

[223]R. Frischknecht et al., *Ökoinventare für Energie systeme*, Eidgenössische Technische Hochschule, Zurich, 1995, 1817p

Table 2.16 Primary energy consumption indicator (MJ/unit).

Resource	Unit	LCV	HCV	CED
Oil gas	Nm^3	40.9	45	–
Natural gas	Nm^3	35	39	38.29
Mine gas	Nm^3	35.9	39.8	39.8
Crude oil	kg	42.6	45.6	45.8
Lignite	kg	8	9.5	9.9
Hard coal	kg	18	19	19.1
Natural uranium (contained In hexafluoride)	kg	460000	900000	560000
Potential energy of water	MJ	1	1	1
Forest wood (dry)	t	18500	20300	=HCV

equal to around 278,000 kWh). To produce one kg of 3.5% enriched uranium requires 8.16 kg of natural uranium. It thus takes 7.58 kg of natural uranium to produce 1 TJ$_e$. This corresponds to 0.007 kg of uranium U$_{235}$, or the potentially recoverable energy of 6.82 primary TJ. The HCV of one kg of natural uranium is thus 900,000 MJ. If the uranium U$_{235}$ contained in depleted uranium is excluded, the LCV of one kg of natural uranium is 460,000 MJ. The new version of Ecoinvent uses an indicator called cumulative energy demand (CED), which is close to the primary energy in HCV, but only takes into account the energy that can technically be obtained from a resource: this indicator gives 560,000 MJ for one kg of natural uranium.

The Ecoinvent database considers a primary energy equivalent to the potential energy of water for hydroelectricity, in HCV as well as LCV and CED.

Fuel equivalents

One kg of fuel is the equivalent of a certain quantity of raw energy, according to table 2.16 taken from the Oekoinventare and Ecoinvent[224] databases.

Choosing between non-renewable energy and total energy

As we have seen, the non-renewable character of resources is taken into account in another indicator, based on global reserves of different combustibles and raw materials. The energy criterion should in our opinion include flow-limited renewable energies, e.g. wood energy. Consuming wood or geothermal energy reduces resources for other uses and moderate consumption should be encouraged. On the other hand, producing energy locally (e.g. using photovoltaic modules integrated into a building) does not reduce resources for other users, and should not therefore be counted as consumption of resources in the evaluation. The indicator calculated in EQUER therefore also includes flow-limited renewable energies. The distinction between renewable and non-renewable energy is made in the resource depletion indicator: e.g. the use of wood is

---

[224]Frischknecht R., Jungbluth N., Althaus H.-J., Doka G., Heck T., Hellweg S., Hischier R., Nemecek T., Rebitzer G., Spielmann and M.."Overview and Methodology", ecoinvent report No. 1, Swiss Centre for Life Cycle Inventories, Dubendorf, Switzerland, 2004

included in the primary energy indicator along with other fuels (gas, oil), but not in the resource depletion indicator.

An idea of size

The human metabolism's average energy consumption is 1000 kWh per year, in the form of food (for a weight of 70 kg), i.e. 2400 kcal per day (or 2400 Cal, 1 Cal = 1 kcal = 4180 J). Each of us uses on average 11 times more energy to heat ourselves, 8 times more primary energy for our electricity consumption (taking an overall yield of 30%), 10 times more for industry and 11 times more for transport.

Heat released into the environment can also be counted, but the primary energy consumption indicator appears to be sufficient to include systems' energy efficiency. That said, thermal emissions such as those from air conditioning systems can generate urban heat islands, which increase the air-conditioning load. They should therefore be included in urban-scale analyses.

2.4.3   Water consumption

Another resource indicator is related to water, which is available in limited quantities in some regions. For instance, as a result of climate change, supply may be threatened in southern Europe. Different types of resource are included: river water, groundwater (but not seawater, which is supposedly unlimited). An overall indicator, expressed in m^3, can be attained by adding together the various consumptions given in life cycle inventories.

2.4.4   Final waste

As we saw in the preceding chapter, several types of waste exist: inert, banal and dangerous. To make it easier to interpret results, it is useful to reduce the number of indicators. A single indicator has been devised for final waste by weighting different types of waste according to their cost. The storage costs at landfills corresponding to the different classes are given in the table below.

Table 2.17 Final waste generation indicator.

Type of storage centre	Type of waste	Costs, margin (ave.)	Tonne equiv. (class III)
class III	inert waste	10 euro/t[225]	1
class II	banal waste	54 euro/t[226]	5.4
class I	dangerous waste	250 euro/t[227]	25

The indicator defined in EQUER is the sum of the quantities of the different types of waste multiplied by the equivalency factors given in the last column.

---

[225] Waste is presumed to be sorted
[226] ADEME, Le prix de la mise en décharge des déchets non dangereux gérés par les collectivités, March 2006
[227] ADEME, Les marchés des activités liées aux déchets, situation 2005/2006 et perspectives 2007

For radioactive waste, the quantities of the different types of waste (low, average and high activity) have simply been added together with no weighting, the indicator being expressed in dm^3 of radioactive waste (cf. paragraph 2.4.7).

2.4.5 Heavy metals

General toxicity indicators are sometimes supplemented by specific indicators for pollutants of particular concern: carcinogenic substances and heavy metals. The values for the indicators of such substances are given in table 2.18, with the indicator expressed in kg of lead equivalent (Goedkoop et al., 1995)[228].

For cancer risks, toxicity linked to heavy metals is integrated into the DALY indicator defined at §2.3. Other diseases related to heavy metals (e.g. effects on the nervous system and liver) are not included in the current version of the DALY indicator due a lack of sufficient data.

Table 2.18 Indicator of pollution by heavy metals.

Antimony	2
Arsenic	1
Barium	0.14
Borom	0.03
Cadmium	3
Chrome	0.2
Copper	0.005
Lead	1
Magnesium	0.002
Mercury	10
Molybdene	0.14
Nickel	0.5

2.4.6 Carcinogenic substances

The values for the indicators of substances contributing to this theme are given in table 2.19, with the indicator expressed in kg of PAH (polycyclic aromatic hydrocarbon) equivalent (Goedkoop et al., 1995)[228].

Table 2.19 Carcinogenic indicator.

Arsenic	0.044
Benzene, ethylbenzene, C_xH_y aromatics	0.000011
Benzo(a)pyrene, fluoranthene	1
Nickel	0.44
Chrome 6	0.44
Tar	0.000011
PAH	1

The number of carcinogenic substances is in fact higher, and the above list comes from the air quality recommendations published by the World Health Organization (Goedkoop et al., 1995).

---

[228] M. Goedkoop et al., *The Eco-indicator 95, final report*, NOVEM, 1995

Carcinogenic effects are integrated into the DALY indicator defined at §2.3, which, as we have seen, makes an early-stage evaluation to obtain a number of years of life lost.

2.4.7 Radioactive waste

Waste is defined according to its activity and lifespan. Category A waste has low or average activity and is due to be stored for 300 years. Category B waste also has average or low activity, but contains very long-range items, like category C waste, which has very high activity. B and C type waste must remain confined for thousands of years. The indicator used in EQUER is the sum of the quantities of this waste expressed in dm^3.

2.4.8 Other indicators

Other indicators relate to noise or risks. Evaluating sound nuisance by simply adding together emissions expressed in $Pa^2.s$ is difficult, given that the level of noise perceived does not only depend on emissions, but on other phenomena like reverberation. Regarding risks, data are currently too uncertain to create a reliable indicator.

2.5 URBAN ENVIRONMENTAL INDICATORS

Improving environmental quality in the building sector requires good integration into the site, taking outside conditions into account, in particular urban choices and the local environment. For example, in a town that recycles its waste, a service area must be sufficiently large to be able to sort waste.

The choice of building site can also have significant consequences on future occupants' transport requirements, and thus on energy consumption and associated emissions. Decision-makers would do well to consider existing or planned public transport networks and to define a general layout plan: how the building is orientated in relation to the south will have consequences on heating requirements, and perhaps air conditioning. Grouping buildings together reduces heat loss, the need for outside lighting, and people's consumption of transport, goods and waste. Lastly, it is important to choose a site that is protected from pollution from industrial installations (depending on dominant winds) or transport infrastructures. This concerns emissions in the air, but also noise. These examples illllustrate the importance of cooperation between contractors, designers and spatial planning decision-makers.

For urban sites, two types of tool have been developed to reduce environmental impacts in towns: indicators and scorecards. Indicators supply information, scorecards facilitate action. The environmental indicators for a town can be used to identify problems and measure the efficiency of proposed solutions.

In rural areas, some criteria, like air quality, are generally less important, but impacts like greenhouse gases, independent from the place of emission, also concern small towns. The analysis described below can therefore also be interesting for rural areas, even though specific measures will need to be put in place.

2.5.1 Emission source indicators (pressures)

2.5.1.1 Energy flows

Inflows, by type of energy (in particular with the percentage of renewable energy produced locally), and flows consumed by sector, give an overall image that can be supplemented by a detailed analysis concerning district heating (market share or % of buildings connected, overall yield, number of days stopped during a heating season), local energy production (relative share, competitiveness), consumption per sector (performance of heating systems in kWh/m^2 for residential, tertiary, school sectors, etc.), energy intensity in the industrial sector, the consumption of the town's operations (public buildings, street lighting, etc.), energy consumed by public transport (relationship of its consumption to total transport sector consumption), and the energy performance of transport (consumption per km and per passenger).

Urban decisions (restrictions on architectural design, orientation of buildings and mutual masks, etc.) have significant consequences on energy consumption, which can be evaluated by the % of buildings exposed to the sun.

Regarding street lighting, the total power and number of light sources per km of street indicate the light density. Performance can be evaluated from the annual kWh consumption of a light source. Buildings constructed in peri-urban zones often generate additional expenditure on street lighting.

The annual consumption of non-renewable energy per inhabitant can provide interesting information. The demand for peak electricity in W per m^2 of useful building surface indicates the efficiency of electricity demand management policies.

2.5.1.2 Transportation

This sector involves factors such as: the number of km (or hectares) of roads per inhabitant; the number of km travelled per year using public transport; the total length of public transport networks per inhabitant and the disposition of different modes of transport; the rate of motorization per household; the average distance between home and work, and possibly home and shopping; the number of km covered in private cars per year per inhabitant; the % of buildings located less than 300 metres from shops; and the % of road surface assigned to pedestrians and cyclists. The density of traffic on a given road is expressed as the number of vehicles per hour. This value is around 500 for an average street in the centre of town, and can be as much as 2000 to 3000 on a motorway.

In Switzerland, each inhabitant makes on average 500 journeys per year in public transport, compared to 150 in France[229]. 50% of journeys in private cars cover a

---

[229]Urban engineering, April 1998

distance of less than 3 km, and 10% less than 500 metres. In the Netherlands, one third of trips are made on a bicycle.

The densification of a town can reduce the need for transport, but it has other limitations: inhospitable urban planning leads to high levels of weekend and holiday departures. High buildings also require lifts, and vertical transportation consumes eight times more energy than horizontal transportation[230].

Transport databases are not always homogenous, and often do not include infrastructure (e.g. construction and maintenance of roads, bridges, vehicles, replacement of tyres, etc.). In addition, emissions of new vehicles are on a downward trend: recent private car models claim greenhouse gas emissions of 120 g CO_2/km, while the Ecoinvent database gives a value of around 320 (including infrastructures). The average occupancy of a vehicle is about 1.6 people. A useful functional unit is people per km (pkm), i.e. the fact of transporting one person for one km. For private cars, emissions are thus 200 g CO_2/pkm, compared to 110 g CO_2/pkm for buses.

2.5.1.3 The water cycle

A network's performance is expressed by its yield (relationship between the volume distributed and the volume produced). This is generally around 80%, which means that leaks represent 20% of the quantities produced. User behaviour can be generally represented by the average consumption per inhabitant, in total or differentiating the residential sector. The sewer connection rate evaluates the performance of the sanitation network. Other useful criteria include the % of reused water, the rainwater recuperation rate, and the % of waterproofed ground.

2.5.1.4 Waste management

Managing household waste is one of the biggest tasks in densely populated zones. Waste generates soil, air and water pollution and dealing with it is costly. Methane recuperation and incineration are sometimes practised. Incinerating household waste brings other problems, in particular the emission of toxic gas into the air (e.g. dioxins), but significant progress has been made in terms of pollution control and energy recovery.

Useful prevention measures include reducing the tonnage of waste produced and setting up sorting, collection and recycling for waste matter.

The indicators used are the weight of waste per inhabitant per year, and the tonnage of types of recycled matter: paper, glass, metal, batteries, oil, textiles, CFCs. Compost is also an option for organic waste.

The description of the treatment structure (landfill, incineration, compost) can be supplemented by a more detailed analysis in case of heat recovery or cogeneration. Energy production must then be evaluated (e.g. taking the quantity of thermal and electric kWh per inhabitant per year).

---

[230]The average energy used per kilometre is 0.29 kWh/km for horizontal urban transport horizontal and 2.32 kWh/km for vertical transport, according to Dr S.J. Marvin (Newcastle University).

2.5.2 State indicators

2.5.2.1 Air quality

Authorities closely monitor air pollutants that can impact on people's health, and in some towns SO_2, NO_2, CO, ozone, hydrocarbons and particles are measured. Indicators are then defined, like the number of days when the concentration of pollutants exceeded an alert threshold, along with indexes that aggregate information on different pollutants, measuring stations and time scales[231] (averages over 1 hour, 8 h, 24 h, etc.).

2.5.2.2 Contribution to the greenhouse effect

CO_2, CH_4 and other gases emitted by various urban activities can be measured, using the equivalence rules presented in the previous chapter. The most common method in this domain is probably carbon assessment, but a lifecycle assessment would provide a multi-criteria evaluation.

2.5.2.3 Noise

Appropriate traffic management can decrease sound emissions. Reducing speed limits and organizing traffic lights can control flows. Impacts can be reduced to some extent by planting trees and creating a distance from building facades (although plants are not particularly efficient in reducing low frequencies). It is possible to map the sound levels in a town, and then set up localized protection (such as anti-noise barriers). Several types of indicator can be defined, e.g. the number of inhabitants exposed to a sound level above a threshold value, or the number of complaints made about noise.

2.5.2.4 Land use and green space

The issue of land conservation cannot simply be resolved by making buildings higher, since "towers' bring other environmental and social problems. Good spatial management requires careful planning, involving in particular: separating residential and industrial zones; the proportion of leisure areas; matching roads with transport requirements; managing urban expansion; and protecting "natural" and rural zones.

The most commonly used indicators are the proportion of built-up areas (or pristine areas), their division by usage (residential, commercial, mixed, leisure, roads), and the transformation of space (from natural to farmed, from farmed to built-up, etc.). The built-up surface per inhabitant also provides information on the consumption of space. In general, the residential sector consumes around 60% of built-up space. Commercial and mixed sectors and roads share the remaining 40% in equal proportions.

Non-built-up areas represent around 30% of the total urban area, but this value can vary widely from one town to another. The proportion of non-built-up space does not necessarily provide a good indicator of the density of green space because the urban

---

[231] Javier Garcia et al., Les indices de qualité de l'air, Presses de l'Ecole des Mines de Paris, September 2001

zone considered may encompass villages separated by rural zones. Non-built-up areas can be sub-divided into forests, parks and sports fields, natural reserves, watercourses and agricultural areas.

Much emphasis is often laid on the surface area of parks and sports fields per inhabitant. Natural reserves, especially wet zones and marshland, are vital to preserve threatened species and biological diversity. It may therefore be worth evaluating the surface area of natural aquatic areas (e.g. lakes and rivers).

The quality and quantity of "natural" areas condition their biodiversity, which can be estimated by the number of animal and vegetable species present in a town. This number decreases when spaces are divided up: for a surface area equal to S, the number of species is around twice as high in an area of surface S than in two areas of surface S/2.

Indicators have been defined to characterize the transformation of space. This involves establishing the typology of an area, e.g.[232]: continuous urban zone, discontinuous urban zone, urban green areas, industrial zone, roads and railways, conventional arable land, organic crops, intensive pastures, forests, etc. An indication of natural degradation is then defined for each type of area (e.g. 95% for continuous urban zones, 10% for organic crops). The indicator of natural degradation due to land transformation is then obtained by multiplying the surface area of the terrain by the difference between the degradation indicators after and before the transformation and by the duration of occupancy of the terrain.

Another indicator that is sometimes used is the ecological footprint, developed at the University of British Columbia[233] (Canada). This represents the area of land required to supply a population with goods and to manage waste (in particular farming land for food and other produce, and land to collect drinking water). This indicator can be calculated at the scale of a town, a country, the world or even an individual[234]. For instance, France's average ecological footprint is around 5 ha per capita[235], whereas the available surface area is only 3 ha per capita. For the world in general, the ecological footprint is 2.3 ha/person for an availability of 1.9 ha/person. Thus, it would take 2.8 times the surface of the planet to provide all inhabitants with the standard of living of the average French person, and 5 planets to provide the same standard as in the USA (9.6 ha/person). The ecological footprint for the town of Toronto can be broken down into 32% food requirements, 24% transport, 21% buildings, 18% other products and 5% waste management.

This indicator is useful for raising awareness amongst the general public; a surface area of land is easier to visualize than a more abstract indicator. However, it does not

[232] Guinée J. B., (final editor), Gorrée M., Heijungs R., Huppes G., Kleijn R., de Koning A., van Oers L., Wegener Sleeswijk A., Suh S., Udo de Haes H. A., de Bruijn H., van Duin R., Huijbregts M. A. J., Lindeijer E., Roorda A. A. H., Weidema B. P.: Life cycle assessment; An operational guide to the ISO standards; Ministry of Housing, Spatial Planning and Environment (VROM) and Centre of Environmental Science (CML), Den Haag and Leiden, The Netherlands, 2001, 704 p.
[233] William Rees and Mathis Wackernagel, *Our ecological footprint: reducing human impact on the earth*, New Society Publishers, Gabriola Island, Canada, 1996
[234] cf. http://www.earthday.org
[235] http://www.footprintnetwork.org/images/article_uploads/France.pdf

integrate all aspects of environmental quality, in particular toxicity and ecotoxicity. For example, the method does not include dioxins, PCBs, heavy metals or plutonium[236]. It would therefore be dangerous to base decisions on this criterion alone, and the same can be said for other single criterion approaches, such as carbon assessments. Two North American architects used this method to evaluate a building, comparing its ecological footprint to the surface of the parcel of land containing it[237]. According to this principle, an individual or organization could pollute more when it possesses a large terrain, which is socially debateable. This contribution illustrates the question of ratios, which should be clearly analyzed to understand which groups are likely to be favoured by the choice of a performance indicator.

2.5.2.5 Soil contamination from heavy metals

The concentration of lead and cadmium can be measured on samples taken from critical zones (e.g. close to roads with heavy traffic or landfills) and protected zones (to obtain a "baseline" concentration). The indicative thresholds are 50 mg/kg for lead and 1 mg/kg for cadmium[238].

Effective control entails making a list of landfills and industrial sites that present a danger of soil contamination.

2.5.2.6 Water quality

Waterproofing caused by built-on land and asphalted roads hinders the regeneration of water tables underneath towns. Industrial activities and landfills are the main causes of water pollution, due to heavy metals (cf. paragraph 2.4.2.5) and organic compounds. Another cause of pollution is neighbouring farming land, where the use of fertilizer, pesticide and herbicide has significant consequences (cf. paragraph 1.2.2). The different sources of pollution can render underground water unfit for consumption and make treatment necessary.

In general, only aquifers used for water supply are monitored. In some countries, samples are more systematic.

Surface waters (e.g. rivers) were for a long time used as sewers, although this practice is disappearing thanks to water treatment plants. Not all wastewater sources are linked up to the sewerage system. The water quality of a river must be measured at several points when there are significant risks or several confluences. If no polluting discharge is made close to the town, it is sufficient to choose a representative point upstream and downstream of the treatment plant. The measurements at each point generally monitor the concentration of nitrates (NO_3), ammonia (NH_4-N), oxygen and phosphates. The biological demand for oxygen is sometimes also used as an indicator.

Drinking water supply must meet with European Union criteria. Regular samples must be withdrawn to measure the levels of nitrates, sulphates (SO_4), heavy metals,

---

[236] Frédéric Paul Piguet et al., L'empreinte écologique : un indicateur ambigü, revue Futuribles No. 334, October 2007
[237] Victor Olgyay and Julee Herdt, The application of ecosystems services criteria for green building assessment, Solar Energy Volume 77, Issue 4, October 2004, Pages 389–398
[238] Seminar *Urban energy and environmental indicators*, Rennes, February 1994

pesticides, etc. (French regulations establish 64 parameters for drinking water)[239]. The hardness of water, expressed in hydrotimetric degrees (°df, content of calcium salts and magnesium, bicarbonates and other salts), conditions the quantity of detergent required for washing. The higher the level, the more detergent is needed and the greater the downstream impact (although detergents are increasingly more biodegradable). To prepare and distribute one m^3 of drinking water requires 10 g of chemical products and from 0.5 to 1 kWh[240], for a given technology.

Pumping underground water must allow the water table to regenerate, otherwise the balance will be upset. The level of aquifers therefore needs to be monitored. Demand is around 100 to 150 litres per day per person in the residential sector. Taking all uses together (commercial, industrial, etc.), this value can be as high as 250 litres. The choice of equipment (see chapter 4) can be effective if coupled with awareness-raising and information for occupants.

Water reserves are often insufficient in towns themselves, and water must be diverted from rural zones. However, the quality of this water is mediocre due to the impact of farming activities and diversion distances are tending to increase (on average 30 km but sometimes over 100 km). These high distances risk diminishing the network's yield, which sometimes drops to 70%, even 50%. These limits make it necessary to introduce policies to reduce consumption and pollution. During critical periods, bans may be necessary (e.g. on washing cars, watering gardens).

High water consumption means more wastewater treatment. Urban connection rates are also an indication of quality. The many causes of pollution (detergents, phosphates, industrial effluent) often render current systems insufficient, and new techniques are progressively being implemented. The quality of treatment can be evaluated by the chemical and biological oxygen demand (COD and BOD), the level of nitrates (limited to 40 mg/l in France, and 30 mg/l in Germany) and phosphates (limited to 1 mg/l in France, and 2 mg/l in Germany). COD is limited to 120 mg/l in France (130 in Germany), and BOD to 40 mg/l in France (30 in Germany). The separate collection of rainwater reduces the flows to be treated in a proportion ranging from 10% to 20%, and in particular decreases peak flows.

The water purification process, which uses mechanical, chemical and biological (bacteriological) processes, produces purified water and sewage sludge. This sludge contains around 80% water and is dried. It then contains over 80% organic matter along with nutritive elements (nitrogen, phosphorus and potassium), and can be used as fertilizer. However, it also contains heavy metals like lead, cadmium and mercury, coming from gutters for example, or industrial processes (like galvanization). Consequently, soil fertilized using sewage sludge must be monitored, and this practice is increasingly banned in some countries, like Germany. In France, the concentration limit is 300 mg/kg for lead, 8 mg/kg for mercury and 15 mg/kg for cadmium. Sludge that is not used for fertilization is considered as waste.

---

[239] S. and P. Déoux, *L'écologie c'est la santé*, Ed. Frison-Roche, Paris, 1993
[240] *Manuel d'écologie urbaine et domestique*, Ed. Le Vent du Chemin, St Denis, 1992

In conclusion, indicators can be precisely defined, with a description of the associated calculation method and an explanation of the unit of measurement, the recommended update frequency, the initial data producer, the disseminating organization and the population concerned.

Geographic information systems facilitate the mapping and spatial representation of environmental indicators. They can also provide statistical data that can be used to evaluate the potential benefits of the envisaged measures.

Scorecards group indicators by theme. They can be used to monitor a situation's progress, assess the efficiency of decisions taken (response indicators) and draw up action plans based on measured results. Infrared thermography on an urban scale (e.g. aerial views taken by helicopter) can be used to raise awareness among the public, without necessarily resulting in a quantified indicator.

Indicator systems that provide a global evaluation have been defined, for example:

- the set of environmental indicators used by the OECD[241], organized into pressure, state and response indicators;
- the indicator systems defined by the European Environment Agency[242];
- systems more specifically related to the building and urban planning sectors (the European research network CRISP[243], the European project RESPECT[244]).

2.6   ENVIRONMENTAL INDICATORS AND SUSTAINABLE DEVELOPMENT

The concept of sustainable development is often used for commercial ends, but it responds to a genuine issue by questioning current development. It constitutes a real challenge because of its inter-disciplinary nature: a client with a programme to "construct a building that meets the needs of the present without compromising the ability of future generations to meet their own needs" would not make it easy for project managers. The objective is so general that it is difficult to apply in practice. One suggestion is to break it down into sub-issues, with each sub-issue having its own objectives according to specific criteria.

This kind of breakdown is arbitrary. Different ways of organizing the breakdown are currently proposed[245,246,247]. Three major aspects stand out: ecological, economic and socio-cultural. For each of these, a list of criteria is given in table 2.20 as an example.

Environmental indicators can also come under a broader grid that includes economic and socio-cultural aspects. The disadvantage is that evaluations are more

---

[241] *Towards sustainable development–environmental indicators*, OECD, March 1998

[242] http://www.eea.europa.eu/

[243] cf. http://crisp-fp7.eu

[244] cf. http://www.respectproject.eu

[245] European ECO-HOUSING project, Environmental co-housing in Europe, European project Number: NNE5/2001/551, final report, January 2006, 29p

[246] Club Bâtiville, *Construire: quelques enjeux de demain*, Cahiers du CSTB No. 3179, December 1999

[247] European LEnSE project, http://ec.europa.eu/research/fp6/ssp/lense_en.htm, 2007

Table 2.20 Sustainable development criteria for buildings, example of grid.

Domain	Criteria
	Ecological Resources (energy, water, raw materials, soil)
	Emissions in air and water (greenhouse gases,
	damage to ozone layer, acidification, eutrophication,
	toxicity to humans, fauna and flora)
	Waste, including radioactive waste
Economic	Investment
	Operation
	Servicing
	Maintenance
	Dismantling
	Sustainability, heritage value
Socio-cultural	Serviceability, adaptability
	Comfort (visual, thermal, acoustic, olfactory)
	Health (cancer, other diseases, accidents)
	Image
	Personal development and employment
	Interface with networks

subjective, in particular in the socio-cultural domain and in weighting between criteria, along with a degree of arbitrariness in the economic domain (the cost of the same technology can vary widely from one region to another, and one project manager to the next). The advantage is that it is more compatible with decision-making processes, which necessarily take the diversity of criteria into account.

A process to improve quality is not fixed: it is a praxis, in the sense of a practice that respects the autonomy of the actors concerned, and is therefore bound to transform and bring out new knowledge during its implementation[248]. One possibility is therefore to devise the type of grid presented above with the actors concerned. This approach fits in with the idea of "governance" that is increasingly a component of sustainable development, and can also be linked to the principle of subsidiarity: i.e. making each theme correspond with the appropriate decision level (e.g. global warming with planet level, acid rain with continental level, waste water management with municipal level) while respecting the interests and points of view of the individuals and groups concerned.

2.7 CONCLUSIONS OF CHAPTER 2

Environmental indicators and indicator systems have been devised. Indicators are a way of evaluating, usually in a relative rather than absolute way, the environmental consequences of decisions. Different choices can thus be compared to select options

---

[248] cf. C. Castoriadis, *L'institution imaginaire de la société*, Ed. du Seuil, 1975

that minimize potential environmental impacts. This type of evaluation is not absolute, since the very complex, perhaps infinite, chain of cause and effect remains largely unknown (cf. chapter 1): greenhouse gas emissions lead to global warming, but current models cannot predict local consequences (e.g. storms and floods) whose impacts (i.e. victims, destruction of buildings) generate new activities (reconstruction) and so new environmental impacts. It is therefore not currently possible to evaluate environmental impacts in an absolute way, and it thus seems reasonable to apply a principle of precaution. Environmental indicators, by allowing us to evaluate potential effects, make it easier to take this principle into account in the decision-making process.

Indicators should not be used in isolation: together they form indicator systems. Marketing measures often tend to isolate one indicator that is favourable to a product or industry. It is obviously important to remain vigilant and demand information based on a consistent, comprehensive set of indicators, e.g. the system defined in the EQUER method (cf. next chapter) or the AFNOR standard NF P01-010[249].

The priority put on some environmental themes, and thus on the corresponding indicators, depends on international commitments (e.g. protecting the climate and the ozone layer), regional regulations (e.g. air quality in towns, water resource management) and even individual choices (e.g. comfort). One of the challenges of environmental management is to conciliate these different levels in a decision-making process that involves all stakeholders.

---

[249] AFNOR, standard NF P01-010, Qualité environnementale des produits de construction–Déclaration environnementale et sanitaire des produits de construction, December 2004

Chapter 3

Methodologies and professional tools

Environmental quality is the fruit of teamwork, starting at the first phases of a building project. The initial choices (i.e. whether to build or renovate, the site, rough layout) determine future performance over the building's lifetime. Since this lifetime is long, impacts generated during the utilization stage are generally high in comparison to one-off impacts during construction. This chapter therefore includes recommendations to carefully design buildings to reduce the flows generated during utilization: water and energy consumption (for heating, possibly air conditioning, lighting, hot water, ventilation and other electricity uses such as domestic appliances or computer equipment), transport, waste production, etc.

This kind of approach involves thinking in terms of overall cost (including both investment and operation) and not just in terms of investment costs. The notion of overall environmental cost includes "external costs", i.e. environmental costs borne by the community (impacts on health, the planet, etc.). This kind of optimization in the community interest will require regulations (e.g. thermal or acoustic regulations, banning CFCs), taxation (e.g. tax on energy consumption or greenhouse gas emissions), or positive action from those involved, possibly supported by labels (e.g. the Qualitel label, solar and low-consumption building labels, etc.).

High environmental quality (HEQ) certification does not necessarily represent actual commitment in terms of performance. For example, the thermal quality level is low compared to the equivalent in Switzerland, Minergie Eco[250], and some "HEQ" buildings consume more than the Ecole des Mines, which is an old building without insulation. Environmental performance is not really covered: the 14 "targets" do not include aspects like climate change or depletion of resources. The split between energy, materials, water and waste does not make it possible to attain overall optima: e.g. the thickness of insulation concerns both materials and energy, and a CO_2 equivalent criterion could be used to optimize emissions linked to the entire life cycle (i.e. manufacture, use and end of life), which is not possible with separate evaluations. Studies are therefore being undertaken to improve performance, both at a theoretical level for evaluation methods, and at a practical level for technical innovations and experimental projects.

---

[250]www.minergie.fr

Two main tools have been developed in the environment domain: life cycle assessment (LCA) and environmental performance assessment (EPA). EPAs apply to organizations, and in particular methods to assess the consequences on the environment and people's health of a source of pollution (e.g. factory, electricity plant) whose location is known. This method is not suitable for an overall study of a building, where pollution sources are multiple and not always easy to locate: during the design stage, an architect does not generally know where the materials used by builders will be produced. On the other hand, it could be used to analyze a particular site manufacturing materials or components. It could also be used to evaluate health impacts on the users of a building, and emissions released from that same building (e.g. coatings, equipment, activities), but this type of tool does not currently exist.

Life cycle assessments (LCAs) study the impacts on the environment (i.e. physical environments, natural resources and living beings) of a system comprising all activities associated with a product, from the extraction of raw materials through to the elimination of waste. It can therefore be used to quantify all of the physical exchanges between a system and its environment, assess its contribution to different effects on the environment, and stipulate and set out options to reduce impacts.

LCA can thus be useful when preparing and evaluating a decision in terms of its impacts, for private or public usage. In addition, with its detailed results, a life cycle assessment has modelling possibilities that can be used to look for improvements in the overall balance sheet, taking into account, among other things, technical, economic and social assessments. This method has been used on buildings to develop architectural design tools.

In this publication, we present the EQUER software programme as an example of a building life cycle assessment, along with different technical tools, in particular dynamic thermal simulation and lighting calculations. A "bioclimatic" concept ensures "biological" requirements (i.e. comfort and health) by taking advantage of climate to reduce energy needs, and thus the corresponding impacts. The COMFIE software programme is devised to help implement these principles by anticipating a building's thermal behaviour. This approach is complemented by taking account of additional aspects (such as water, waste and materials), using a life cycle assessment.

After presenting these basic tools, we provide several tips on tackling environmental quality at the various phases of a project (programming, design, worksite and management). The building sector is very diverse, with 29,000 architects, 20,000 engineers working in technical consultancies, 2500 construction economists and 1.5 million people working for building contractors (of whom 100,000 are temporary). This partly explains the difficulty in disseminating knowledge and developing practices: in small outfits, tools and methods need to be simple, which is in contradiction with the wide range of building projects and the complexity of environmental analysis.

General approaches have been developed to tackle overall environmental quality, but the desire to simplify each technical evaluation renders these approaches rather superficial. Some examples are given at the end of this chapter, but the more precise linking of tools in a geometric description of buildings (CAD: computer-aided design) provides a more satisfying prospect for the future. The following diagram illustrates this approach, and gives an overall framework of the various contributions presented in this chapter.

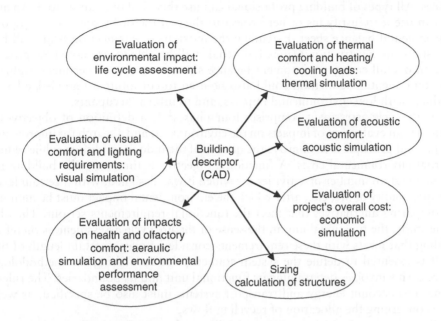

Figure 3.1 Links between environmental assessment and other technical domains.

3.1 LIFE CYCLE ASSESSMENT

Life cycle assessments are most frequently used at international level to evaluate environmental impacts. They involve studying a product from its manufacture, taking into account its components and therefore going right back to the resources taken from the environment, up to the end of its life, including the final waste generated, and covering every stage of its usage. The first applications date from the late 1960s (for fizzy drink containers).

This method involves evaluating the quantifiable aspects of environmental quality, and therefore does not tackle more qualitative aspects like damage to the landscape. Life cycle assessments were initially devised for industrial products. They should therefore be used with care for the building sector: each building is usually unique and maintains close links with the site on which it is integrated and with its occupants. In addition, some evaluations, like human toxicity, are still very imprecise. This currently appears to be the method with the least possible ideology. It can help improve the precision of the more simplified tools proposed to professionals. The EQUER method was put together on these bases by a team initially comprising two research centres: the Department of Energy and Processes at the Ecole des Mines de Paris, and INERIS (Institut d'Évaluation des Risques Industriels), along with professionals from DUMEZ-GTM (currently in the VINCI group), the architect agency S'PACE and Pierre Diaz Pedregal Consultant.

The aim of the EQUER project is to develop a tool for assessing the environmental quality of buildings, so as to help stakeholders apprehend the consequences of their choices. All types of building professional can use this kind of analysis tool. An architect can use it to justify his or her project to the contractor by presenting a rigorous environmental balance sheet. Extending the missions of engineering firms will boost their status in construction teams. Industrials will be able to promote new products, since an overall energy-environment balance sheet constitutes a pertinent evaluation of a component. The method could also be used to constitute a knowledge base on buildings with low environmental impacts, and to inform occupants.

A life cycle assessment comprises four phases[251]: a definition of objectives, an inventory, an evaluation of impacts on the environment, and a search for improvement. The general principle is to improve the quality of the indoor conditions while reducing external environmental costs. A functional objective is thus set – the building must permit a certain number of activities for which it was designed, with a certain level of comfort, quality of life, etc., after which the environmental impact must be minimized by comparing alternatives that meet the functional requirements set out. The object of the study, the functional unit in the sense of the life cycle assessment, is therefore a building that meets with these requirements considered over a certain length of time.

It is essential to define the system studied prior to making the methodological choices. This involves determining its functional unit and its boundaries. The rules for taking into account energy and transport systems must also be specified, as well as those concerning the allocation of recycling flows.

The next stage is the inventory, which lists and quantifies the flows of matter and energy going in and out of the defined system. The environmental impacts are then evaluated by aggregation (i.e. calculating the indicators from elementary flows, cf. preceding chapter) and lastly, improvements can be sought.

Due to the multiple sectors of application, life cycle assessments must be adapted to each study in line with specific objectives. This requires a great deal of transparency in the methodological choices made (definition of problem raised, functions studied, etc.) in order to make a pertinent interpretation of the results.

Several examples of how to use an LCA in the building sector are given in paragraph 3.1.7 and in chapter 5.

3.1.1   Methodology used

The EQUER project involves developing a simplified simulation tool for modelling a building's construction, utilization, renewal of components and deconstruction, including any re-use and recycling. A computer tool is developed to facilitate comparison of the alternatives and so help decision-making. Calculations are based on digital simulation to represent reality more precisely. Chaining was accomplished with a thermal simulation tool, COMFIE[252] using an approach resulting from research on computer data exchange (STEP[253] standard). This establishes a link between energy

---

[251] AFNOR, norme X 30–300, *Analyse du cycle de vie*, March 1994
[252] Bruno Peuportier and Isabelle Blanc-Sommereux, *Simulation tool with its expert interface for the thermal design of multizone buildings*, International Journal of Solar Energy, 1990 vol. 8 pp. 109–120
[253] Bo-Christer Björk and Jeff Wix, *An introduction to STEP*, VTT (Technical research centre of Finland) and Wix McLelland Ltd, 1991

analysis and environmental analysis. Thus, energy is no longer perceived as a simple quantity of kWh, but viewed according to a series of environmental criteria. In addition, energy does not only concern heating or lighting: energy recovered in an incinerator coupled to a heat network is taken into account, as is the energy required to manufacture construction materials, the transport generated by the building, and the water supply. Lastly, aspects not linked to energy are taken into account (e.g. water management, construction materials). The scope of analysis is therefore significantly wider, allowing for more comprehensive studies, several examples of which are given below.

Definition of a functional unit

The system considered is an entire building, which is comfortable and sound, in a given region, over a fixed length of time based on a given usage (accommodation, tertiary, etc.), and with a certain number of occupants. The duration of the analysis can be set by the user depending on the context. A default value of 80 years is proposed. The impacts evaluated can then be related to $1\,m^2$ of building, which makes it easier to compare with standards or best practices.

System boundaries

Along with the building, it is possible to include impacts caused by water and energy supply, water treatment, and waste, as well as the transport of materials and people. In addition, energy production and water treatment facilities can be taken into account, since these procedures can be located in the building itself or on its parcel of land, unlike transport and waste processing infrastructures. The system boundaries depend on the target of the study. If the objective is to compare different sites for a new construction, then it should include transportation (e.g. from home to work), household waste management, energy networks (electricity, gas, possibly heat) and water. However, if the aim is to aid design on a plot of land that has already been chosen, then the study will be restricted to the envelope and the building's equipment; the transportation of people can be left out if all of the alternatives compared are equivalent from this point of view.

Environmental balance sheet and profile

The calculation results in a quantification of elementary flows over the building's life cycle, thus determining the inventory corresponding to the functional unit defined above. The final outputs are presented in the form of an environmental profile, i.e. a set of indicators from those presented in chapter 2.

3.1.2   Modeling buildings: An object-based approach

There are many ways to describe a building. A working group at the ISO studied a general modelling of buildings as part of the STEP standard. Some CAD developers propose object-focused models that make it easier to link up with technical evaluation tools, in particular the International Alliance for Interoperability (now called "Building

SMART") with its IFCs ("Industry Foundation Classes[254]"). It is preferable to continue these developments so that in the future we may be able to link this type of tool with CAD and other technical tools (thermal, acoustic, lighting, structure calculations, etc.). However, given the difficulty of defining a universal model, a model adapted to the LCA[255] has been elaborated, with a distinction between:

- Objects manufactured in a factory: products (e.g. bricks, breeze blocks, etc.) and components (e.g. bay windows, sandwich panels);
- Objects constructed on-site – sub-systems (e.g. walls, rooms), and the building itself;
- The worksite (construction, renovation or demolition).

Each object is associated with processes (transport, set-up, operation and maintenance, deconstruction) and an "inventory" containing all of the associated elementary flows: resources taken from the environment and emissions generated. A precise description of this model is given by Bernd Polster[256].

The overall model thus includes entities corresponding to the building's technical objects (e.g. materials, components), processes (transport, energy processes, use of water, waste management, etc.), the site (ambient air, selective sorting, public transport, etc.) and environmental indicators (inventories, water and energy meters, etc.).

A comprehensive description of the corresponding data structures was drawn up following the lines defined in STEP, in particular the graphic representation EXPRESS (cf. figure below). The objects are grouped into classes. Sub-classes have been defined to allow data and methods to be inherited. The very general "element" class contains the notion of an inventory and some methods to count impacts, common to all entities.

The product sub-class represents the basic constituents, which cannot be broken down into sub-sets. This includes for example basic materials used for masonry, wall coverings, etc.

The component sub-class corresponds to objects manufactured by combining simpler elements in factories or on-site. This includes for example bay windows and solar protections. These objects are defined by their basic constituents and procedures (construction, dismantling, partial recycling, etc.).

The sub-system sub-class represents more complex objects formed from products and/or components, such as inside walls and foundations. The notion of a thermal zone is introduced to make a link with the energy analysis. Each zone is associated with consumption of energy and water and household waste management.

Lastly, the building and the worksite make up two other sub-classes that, along with their constituents, include additional data (land use, sources of noise, specific processes, etc.).

---

[254]Patrice Poyet and Jean-Luc Monceyron, *Les classes d'objets IFCs, finalités et mode d'emploi*, Les Cahiers du CSTB, No. 2986, October 1997

[255]Bruno Peuportier, Bernd Polster and Isabelle Blanc Sommereux, *Development of an object oriented model for the assessment of the environmental quality of buildings*, First International Conference "Buildings and the environment", CIB, Watford, May 1994

[256]Bernd Polster, *Contribution à l'étude de l'impact environnemental des bâtiments par analyse du cycle de vie*, doctoral thesis, Ecole des Mines de Paris, 1995

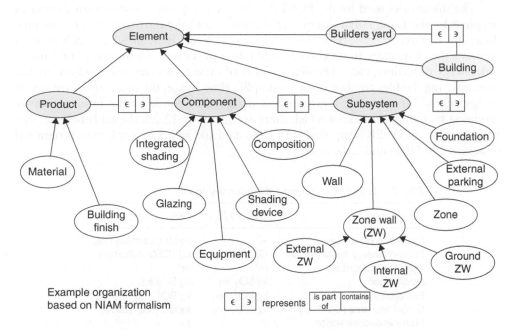

Figure 3.2 Example of a model of a building, organization of data.

This kind of object-based organization can take into account results of studies on the environmental impact of building materials or components (e.g. those done by the Ecole Polytechnique Fédérale de Zürich, the Swiss ministry of the environment BUWAL, etc.).

3.1.3 Data collection

A list has been made of the data available in inventory form for construction materials, energy (electric kWh, thermal kWh coming out of boilers, etc.), transport (tonne-km for lorries, person-km for private cars, etc.) and waste (incinerator with or without heat recovery and/or electricity production). An exhaustive base does not currently exist, but more and more manufacturers make life cycle assessments of their products, and bases are increasingly comprehensive.

Inventories of processes take into account possible early-stage phases: e.g. the inventory of useful kWh supplied by gas boilers considers the extraction, preparation, transport and distribution of gas. The impacts linked to infrastructure are counted if they are significant. For example, transport inventories cover the construction and maintenance of networks, while the impacts of building factories for manufacturing materials and components are considered as negligible. Household waste can be recycled or incinerated, with or without heat recovery. The corresponding inventory is calculated based on the data supplied by the user (yield of the energy recovery, substituted energy, etc.).

The inventories used by the EQUER software programme result from databases supplied by the École Polytechnique de Zürich[257] and the most recent versions produced by Ecoinvent[258,259]. They take the form of Excel files, in which each column contains the elementary flows corresponding to a procedure (manufacture of a material, energy procedure, etc.). The high number of these flows (several hundred) makes manipulating the files cumbersome. To simplify their usage, the inventories are stored in aggregated format. This aggregation is designed to condense the information by grouping them into themes, of which there are currently 12 (cf. the table below). The basic version of EQUER is supplied with 1996 data and an Ecoinvent licence is required to access 2007 data and indicators.

Table 3.1 List of environmental themes considered.

List of themes	Units 1996	Units 2007
Depletion of resources	–	kg Sb (antimony) eq.
Primary energy consumption	GJ	GJ (CED indicator)
Water consumption	m^3	m^3
Acidification	kg SO_2 eq.	kg SO_2 eq.
Eutrophication	kg PO_4^{3-} eq.	kg PO_4^{3-} eq.
Global warming (GWP_{100})	kg CO_2 eq.	kg CO_2 eq. (2007)
Non-radioactive waste	t	t
Radioactive waste	dm^3	dm^3
Odours	m^3	m^3
Aquatic ecotoxicity	m^3	PDF.m^2.an
Human toxicity	kg	DALY
Photochemical ozone (*smog*)	kg C_2H_4 eq.	kg C_2H_4 eq.

Because of the interdependence between industrial sectors, a matrix equation is used to calculate inventories from the Ecoinvent base. For example, to manufacture bricks requires gas and electricity, to produce electricity requires concrete, etc. In addition, each sector calls on general services (e.g. banks, insurance, training, R&D). A matrix resolution is thus needed[260].

Other databases exist, such as INIES[261] devised in France in line with the AFNOR standard P01 010[262]. This base provides FDES (Fiches de déclaration environnementale et sanitaire – health and environment declaration sheets) on numerous construction

[257]Frischknecht R., et al., Oekoinventare von Energiesystemen, 3. Auflage, ETH Zürich/PSI Villigen, 1996

[258]Frischknecht R., Jungbluth N., Althaus H.-J., Doka G., Heck T., Hellweg S., Hischier R., Nemecek T., Rebitzer G., Spielmann M. (2004) "Overview and Methodology", ecoinvent report No. 1, Swiss Centre for Life Cycle Inventories, Dübendorf, Switzerland

[259]2007 version available at www.ecoinvent.ch

[260]Jolliet Olivier, Saadé Myriam and Crettaz Pierre, Analyse du cycle de vie, comprendre et réaliser un écobilan Presses Polytechniques et Universitaires Romandes, Lausanne, 2005

[261]www.inies.fr

[262]AFNOR standard NF P01-010, Qualité environnementale des produits de construction – Déclaration environnementale et sanitaire des produits de construction, December 2004

items, grouped into product families (roads, structure, woodwork, roofing, etc.). However, it does not include processes (production of gas, electricity and water, waste treatment, transport, etc.) and so cannot be used to produce a comprehensive balance sheet. End-of-life impacts are not evaluated, and only the quantities of different types of waste are counted. In addition, some substances are absent from the inventories (e.g. dioxins) and some indicators are over-simplified (critical volumes). Lastly, interactions between sectors are not taken into account in a matrix system. A building-scale extension is in process[263], in line with European[264] and international[265] studies. Similar bases are being developed in the United Kingdom[266] and Finland[267].

Another base is being put together on a European scale (European life cycle data[268]), although it is currently fairly limited and badly documented. It is essential to have access to documents describing systems (i.e. production procedures and end-of-life processing, transportation distances), and setting out hypotheses (resources on several production sites, recycling, etc.) to verify the pertinence of some data: this is also one of the strong points of a base like Ecoinvent, which comprises several thousand pages of documentation.

3.1.4 Software development

The description of a building is done using an existing software programme (COMFIE, muti-zone dynamic thermal simulation, and PLEIADES interface along with a graphical user tool ALCYONE[269]). Then the heating consumption (and possibly air conditioning) calculated by the COMFIE software are transmitted to EQUER, along with all the entry data required for thermal calculations such as the geometry of the envelope and its constitution. Additional variables are requested: water consumption,

---

[263]AFNOR, standard NF P01-020-1, Bâtiment – Qualité environnementale des bâtiments – Partie 1: cadre méthodologique pour la description et la caractérisation des performances environnementales et sanitaires des bâtiments, March 2005
AFNOR, Guide GA P010-020-2, Bâtiment – Qualité environnementale des produits de construction et des bâtiments – Partie 2: guide d'application de la norme NF P01-020-1:2005, September 2007
AFNOR, standard XP P010-020-3, Évaluation de la performance environnementale d'un bâtiment – description du résultat de l'évaluation, de la méthode d'évaluation et de leurs déclinaisons à différentes étapes d'un projet, en cours d'élaboration
[264]CEN/TC 350 /WG1, Workshop on construction sustainability assessment standards: the interface between CEN/TC 350 standards for assessment of environmental performance (product and works level) and standards for products and for construction works – calculation methods under study
[265]ISO, standard ISO/PDTS 21931-1: Framework for methods of assessment of the environmental performance of construction works – Part 1: Buildings
ISO/NP 11368, Building environment design – Energy performance of buildings – Methodology to assess the overall energy and environmental impact by using a holistic (complete) approach procedure, being developed
[266]Environmental profiles data base, www.greenbooklive.com
[267]RT Environmental declaration, https://www.rakennustieto.fi/index/english.html
[268]http://lca.jrc.ec.europa.eu/lcainfohub/datasetArea.vm
[269]cf. http://www.izuba.fr

household waste management (glass and paper sorting, landfill or incineration, possibly with energy recovery) and, if need by the application, transport distances (home to work, home to shops, etc.).

Default values can avoid inputting too many parameters, for example, the desired hot water temperature, the cold water temperature according to the latitude. Some default values are proposed but can be modified by the user (e.g. hot and cold water consumption per person, transport distances to work and shops depending on the site – urban, suburbs, rural, remote – quantity of waste generated per day per person, lifespan of components, etc.). Other simplifications in the interface include assigning the same data to all materials (transport distances, % of waste in construction), or to classes of material (lifespan of woodwork, coatings, other components). However, specific data per component can be used in the calculation.

The EQUER data structure is thus put in a STEP-type file, where (in the research version used) the values of the parameters can be modified if they are different from the values resulting from the simplified interface. From this file, a structure of objects representing the building is automatically created. The inventory calculation can then be carried out using the methods associated with each object, based on the simulation procedures presented in the figure below. In this first phase, methods have been developed for construction and usage, including transport. The building is simulated over the usage phase with a one-year time step. Components and/or their possible constituents are automatically replaced using age meters included in the objects.

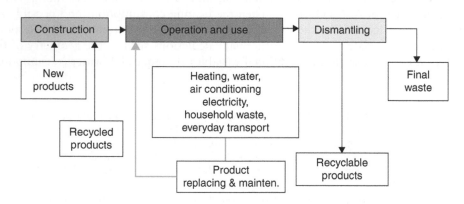

Figure 3.3 Calculation principle of the overall inventory.

The software's output is presented in the form of an eco-profile, with the possibility of visualizing the contribution of each phase (construction-utilization-renovation-demolition) and comparing up to 4 alternatives for one project. The phase called "renovation" corresponds to replacing components at end of life (windows, coatings).

As we have seen, the environmental profile resulting from this evaluation comprises a dozen indicators. Some methods aim to produce an overall environmental performance indicator. One approach consists in weighting emissions according to reduction targets defined by political leaders, e.g. the Swiss ecopoints method, or the Dutch Ecoindicator, which is based on the hypothesis that impacts on humans have a weight of 95%. These approaches all include a degree of subjectivity, and it is currently preferable to continue with a multi-indicator profile that deciders can interpret in line with the context. The priority given to different themes can vary depending on the region (e.g. air quality in very urban zones, water availability, etc.) and stakeholder awareness.

3.1.5 Limitations of the methodology

An LCA only relates to the quantifiable aspects of environmental quality. More subjective considerations of aesthetics or quality of life are therefore not included. Regarding health, the human toxicity indicator is evaluated considering the world's total population and an average dilution of pollutants in the atmosphere, or in the most recent version of the database, a DALY indicator also corresponding to a spatial average. It does not therefore depend on the location of the emission (population density, dominant winds, etc.), although this is generally unknown since designers rarely know in which factory construction materials are manufactured.

Slightly more detailed indicators are proposed (e.g. emission of heavy metals, emission of carcinogenic substances, winter smog, summer smog), but this does not resolve the issue of locating emissions. To evaluate actual impacts (number of respiratory diseases generated or made worse) would involve including indoor air quality in the analysis, which would require other model categories (such as mass transfer in floorings and wall coverings, air movements, etc.).

One limitation that all indicators have in common is imprecise evaluations. It is often difficult to ascertain the margin of uncertainty on data and results, but it can be high. The first level of imprecision concerns evaluating flows of matter and energy (inventory data). The second level concerns the aggregation of effects (potential impacts), e.g. uncertainty regarding the potential for global warming from greenhouse gases other than CO_2 has been estimated as 35% by the IPCC (by chance, CO_2 represents around 80% of the total impact, therefore overall uncertainty is not more than 7%). Lastly, the third level concerns the passage from effects to impacts. This level is not envisaged here beyond the DALY and PDF indicators (cf. chapter 2) because it is not covered by current models. Hopefully, in the future these uncertainties will be reduced.

Another cause of error is linked to the length of analysis periods. For example, it is difficult to anticipate progress in waste treatment techniques, in particular demolitions, which may take place in a hundred years or more. It may be preferable to envisage a probabilistic analysis based on scenarios subject to probability (e.g. 50% chance of recycling a product at end of life).

3.1.6 Example of experimental application at the ÉcoLogis exhibition

The EcoLogis competition regulations included several environmental performance requirements that aimed to encourage, for example, the application of commitments made at the Rio Conference on Environment and Development.

The EQUER software programme, linked to the thermal simulation software, COMFIE, was used to help the winning team in the design. It was an opportunity to test the tool on an experimental project. The winning design sketch (cf. figure below)

Figure 3.4 Design of the house produced for the ÉcoLogis exhibition.

was assessed, then alternatives were proposed, especially for the sunspace design and pre-heating new air.

Recommendations included reducing glass surfaces between the sunspace and the house and using low-emissivity double glazing. Appropriate solar protection was added to the west-facing window, and the surface area of the skylights was reduced to make the most of natural lighting in a corridor. The software did not, however, allow the selection of construction materials based on life cycle assessments. Performance sensitivity to this parameter was still low (except for components that influence flows, in particular energy flows) with regard to uncertainties on data and environmental indicators.

At the end of the design phase, the final project was assessed. The environmental performance of Écologis was taken as a relative value, compared to a reference representing current standards for new buildings in Ile de France (Paris region). This reference house was defined as part of the ATEQUE project (a workshop to evaluate the environmental quality of buildings organized at the ministry of housing by PUCA, Plan urbanisme, construction et architecture), taking into account:

- INSEE (National Institute of Statistics and Economic Studies) statistics on building techniques (type of masonry, joinery, roofing, heating equipment, etc.);
- A representative architectural design (plans supplied by H. Pénicaud).

For the two houses, an average usage scenario was defined: heating at 19°C, air renewal rate of 0.6 volumes per hour, internal gains of 400 W, cold water consumption (and respectively hot water, at 50°C) of 100 litres per person per day (respectively 40 litres), household waste generation of 1 kg per person per day, of which 12% glass and 30% paper. Transport from home to work was not considered.

The results of the thermal calculations show that ÉcoLogis's upper solar opening, super-insulated glazing and ventilation heat exchanger lead to around 20% savings in relation to 1 m² (ÉcoLogis has a surface area greater than the reference). The comparative profile below was obtained considering 50% savings on water flows for sanitary equipment, 7% savings on electricity consumption (reduction of lighting requirements) and 40% paper recycling (60% for glass respectively).

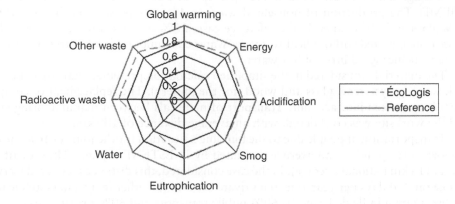

Figure 3.5 Ecoprofile of the ÉcoLogis house based on 1995 standards.

The global warming potential, considering flows only, is therefore reduced by around 20%. Sensitivity analyses showed that the construction phase has a much lower influence than the utilization phase (around 10%), therefore the simplified balance sheet presented above gives a fairly good approximation of the detailed balance sheet. The environmental impact could be reduced further by even greater thermal optimization (cf. the models devised by other prize winners), but the ÉcoLogis presented at La Villette constitutes an initial improvement in relation to the reference.

Regarding indoor conditions, the spaces are well lit, the acoustic protection has been considered, and ventilation assures the air quality. The only problematic aspects are the thermal comfort, due to the lightness of the construction (it was impossible to use heavy materials on the exhibition stand), the presence of west-facing horizontal or vertical windows and the absence of solar protection. A "virtual" house could be designed outside the context of the exhibition with greater thermal mass, well-positioned windows and suitable solar protection.

3.1.7 First sensitivity analyses: Relative contributions of different sources of impact

The evaluation described above was made on a standard detached house, considering statistical averages to represent a typical occupancy scenario. The building considered is a 100 m^2 detached house in Ile de France (Paris region). The single-storey house is made of concrete (16 cm walls, 10 cm slab floor), with outside insulation (8 cm in the walls, 14 cm on the ceiling, 5 cm under the floor) and 10 m^2 of south-facing windows (double glazing). Heating and water are gas-fired. The heating is set at a constant temperature (19°C), resulting in consumption of 8500 kWh/an.

The building analysis duration is set at 80 years. The building is occupied for 300 days per year by a 4-member family including 2 children. Electricity consumption (household appliances and lighting) is 3200 kWh/year, which corresponds to standard equipment. The inventory of electricity kWh (lighting, etc.) corresponds to the 1996 French production split (74% nuclear, 14% hydraulic, 12% thermal).

Average data on household waste (quantity and recycling) were obtained from ADEME. The production of household waste is 1 kg per person per day, of which 12% glass and 30% paper. In the reference case, 20% of glass is recycled. The other waste is incinerated without heat recovery. Water consumption is 140 litres per person per day, including 40 litres of hot water.

The materials considered in the simplified calculation presented here are concrete, rock wool, polystyrene, glass and wood (for the joinery). The replacement of components is not taken into account. Transport of materials between manufacturing sites and the worksite was considered, with an average distance of 100 km.

Transportation of people due to the building's situation for the journey from home to work corresponds to an average evaluated by INESTENE[270], i.e. 12 km for Ile de France (14 km national average). In the case considered, this distance is covered by two people on 230 days per year, one in a private car and the other in a train (statistically, the breakdown in Ile de France is 60% public transport and 30% private cars).

In this standard configuration, it is interesting to evaluate the relative contribution of the different sources of impact. The breakdown is different for each environmental theme, as shown by the figure below.

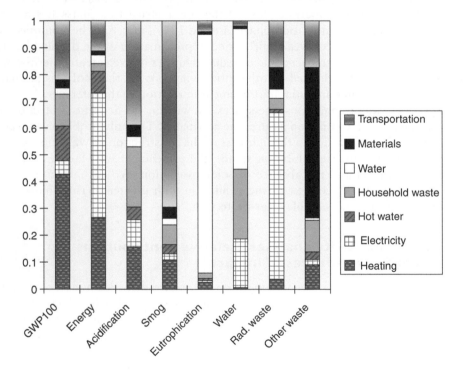

Figure 3.6 Contribution by different sources of impact to the overall balance sheet.

---

[270]Pierre Radanne, Laurence Moulin and Danielle Banneyx, *Aménagement du territoire, mode d'habitat, mobilité quotidienne et environnement*, INESTENE, 1994

In this case, the three main sources of greenhouse gas emissions are waste incineration, heating and transport. For the "primary energy" theme, the highest contribution comes from electricity production. Construction materials only play a significant role when it comes to waste production (when the building is demolished or renovated, and to a lesser extent when it is built).

Ways to improve the ecological balance sheet

After evaluating the performances of this reference, alternatives can be compared to look for ways to improve the balance sheet, with equivalent comfort and quality of life. When it comes to the ecological construction theme, actors often turn towards "natural materials". However, figure 3.5 shows that, in the case of a construction built to recent standards (thermal regulation RT2000), materials only make a low contribution to most emissions. Although some manufacturers have made attempts to reduce pollution (e.g. recycling, use of wood to reduce greenhouse gases), priorities should be focused on more significant sources of impact (these priorities can of course be different in the case of high energy performance buildings). The alternatives considered are thus the following.

Reducing emissions from the home-work journey

It is sometimes possible to choose a place of residence that limits the distance from home to work, and/or gives easier access to the public transport system. Based on the above example and without changing anything else, a home to work distance of 2 km was considered, along with a bus journey for two people (bicycle rides could also be studied, integrating the impact of building cycle lanes).

Reducing heating loads

Thermal regulations have three times imposed a 25% reduction in building heating loads (constant heated surface). However, this only applies to heating in new buildings and the reduction has been compensated by an increase in other energy expenditure, in particular on household appliances.

"Bioclimatic" architectural design can further reduce heating consumption (and lighting to a lesser extent) by making better use of solar gains. External insulation, the use of well-oriented windows (south-east to south-west) and better thermal insulation, pre-heated air in a sunspace or an exchanger are all well developed techniques. These can be supplemented by using energy-saving appliances (e.g. low-energy light bulbs). A saving of around 40% on heating is totally realistic and this was considered in the alternative compared (examples of even greater savings are given in chapter 5). It is worth noting that an apartment in a building equipped with a well-designed heating system consumes less heating energy than a detached house (because outside walls are reduced, and thus energy loss).

Regarding the greenhouse effect, gas heating was compared with electric heating. Electric heating generates peak winter consumption that makes it necessary to use thermal power plants (nuclear plants are more suited to regular operations throughout the year). The yield of a power plant is lower than that of a boiler. According to this calculation, electric heating does not reduce greenhouse gas emissions (even taking gas power plants with a 60% yield, supposing that nuclear power plants are used for at least 4000 hours per year).

Recycling household waste

Recycling household waste is an everyday task that appears relatively easy and realistic. An alternative was therefore defined whereby inhabitants recycle 40% of their paper and cardboard and 60% of their glass. The alternative also supposes that the incinerator fuels a district heating network. Compost was not included in this study, nor was reducing the consumption of packaging, although they could be.

Managing water

Using low-flow sanitary equipment can reduce water consumption by up to 50%. In addition, solar water heaters can supply 40% of energy needs for domestic hot water (and up to 60% in sunny climates). Regarding water pollution (liquid waste, detergents, etc.), data on the impact of sanitation are still too incomplete to be able to compare alternatives on the choice of detergents, for example.

Other alternatives could thus be defined, and the comparison would be different depending on the climate, lifestyle and type of construction chosen. The example given here provides an illustration of the approach. The results are presented in the figure below for the "greenhouse effect" theme.

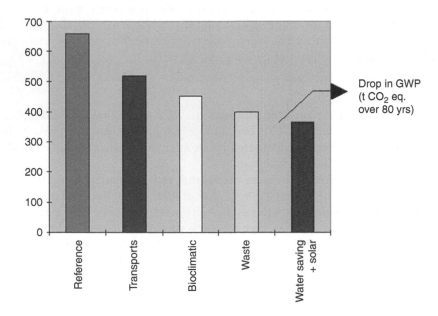

Figure 3.7 Reduction of greenhouse gas emissions.

These results show that a reduction of over 40% in greenhouse gas emissions is possible without reducing the level of comfort in dwellings. More sophisticated techniques could take this approach further still. Demonstration projects have shown the benefits of solar heating systems, for example using transparent insulation (the "Aurore" housing estate in the Ardennes) or direct solar floors (numerous projects in the Alps). The Fraunhofer Institute in Freiburg (Germany) has experimented with a zero energy house, heated by transparent insulation with a back-up system fuelled by hydrogen, produced in the summertime using solar panels. However, energy autonomy is not necessarily the environmental optimum because of the impact of manufacturing systems (especially batteries). Heating networks (geothermal, wood fuel, etc.) offer interesting avenues. Positive energy buildings are presented in chapter 5.

3.1.8 Normalization of an ecoprofile

In the ecoprofiles presented above, the different indicators are represented on a single scale, in relative values from 0 to 1. This makes it possible to avoid major disparities in the sizes of the different results: e.g. for the life cycle of a standard house, the global warming potential is around 600 tonnes of CO_2 equivalent, whereas the resource depletion indicator is around 10^{-8} and aquatic ecotoxity is around 10^7, i.e. a factor of 10^{15} between the extremes. Taking a scale between 0 and 1, the different indicators can be represented on the same diagram. The disadvantage is that each indicator is alloted the same importance, whatever the contribution of the system studied.

Let us take, for example, a photovoltaic installation of 3 kWp (peak kW, i.e. electric power supplied for 1000 W/m^2 of solar radiation and a cell temperature of 25°C) connected to the grid (i.e. without a battery). The aim of the LCA is to compare the impacts of manufacturing this system with the impacts avoided by producing electricity from a renewable source.

To simplify, the stages of the life cycle are:

- reducing silica to obtain silicon;
- purifying silicon;
- moulding ingots and sawing plates;
- manufacturing photovoltaic cells;
- manufacturing modules;
- assembling and integrating them into the building;
- manufacturing the inverter (transforming the direct current produced into alternating current) and electric installation;
- using and maintaining the system;
- disposing of and managing the waste.

Substances taken from or released into the environment are counted for each stage, then an ecoprofile is obtained by aggregating the values according to the previously defined indicators. The results are as follows (see table 3.2 below).

The relationship between the values for the electricity produced and the sum of indicators for the manufacture and dismantling of the system gives what might be called the efficiency of the photovoltaic system over its life cycle. For instance, the primary energy production corresponding to the electricity produced is 8 times the consumption of primary energy for manufacturing and dismantling the system. On the other hand, the system generates almost ten times more waste than the amount

Table 3.2 Results of the life cycle assessment of a photovoltaic system.

Theme	Unit	Manufac.	Elec. produced	End/life
greenhouse effect	kg CO_2	7112	−10557.2	95.54
primary energy	kWh	62167	−486470.4	6.1
acidification	kg SO_2	55.48	−96.36	0.95
smog	kg C_2H_4	2.65	−1.17	0.23
eutrophication	kg PO_4^{3-}	3.73	−5.678	0.03
water	m^3	679	−920.6	3.263
radioactive waste	dm^3	0.47	−5.13	5.87E−06
other waste	kg eq.	3798	−988.4	5553

corresponding to its electricity production, but 10 times less radioactive waste. All of these ratios are given in the table below.

Table 3.3 Balance sheet of a PV system.

Theme	Impact avoided/impact generated
greenhouse effect	1.46
primary energy	7.82
acidification	1.71
smog	0.41
eutrophication	1.51
water	1.35
radioactive waste	10.78
other waste	0.106

In this table, the efficiency of the system is about the same for the greenhouse effect and eutrophication: the impact avoided represents 1.5 times the impact generated. Photovoltaic production reduces the effect of greenhouse gases by around 10,500 kg of CO_2 equivalent and eutrophication by around 5.5 kg of PO_4^{3-} equivalent. Does that mean we should attach the same degree of importance to both of these facts?

In oder to help answering this question, it is proposed to normalize the impacts in relation to average impacts per inhabitant and per year. For instance, one person produces in average in France 8700 kg of CO_2 equivalent and 38 kg of PO_4^{3-} equivalent yearly. The reduction of climate change thanks to the photovoltaic electricity production is then higher than the yearly emission of one inhabitant, on the other hand the reduction of eutrophication is only 0.15 inhabitant-year. The results are shown on the figure hereunder.

This graph is interesting in that it highlights the themes for which the system's contribution is negligible: eutrophication and smog. In this example, the electricity production reduces overall emissions (the equivalent of 10 inhabitant years for primary energy and radioactive waste), except for solid waste (increase of around 0.8 year-inhabitants). So is it beneficial overall? To answer this question, we would need to define a relative weight for the different indicators, for example by attaching as much, or more, importance to the greenhouse effect as to radioactive waste, etc., which implies a degree of subjectivity. It is probably therefore judicious to stop at the stage

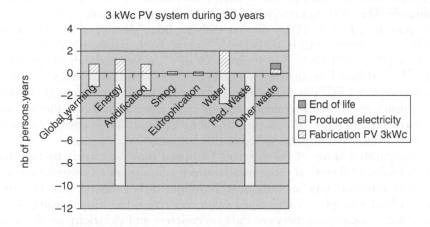

Figure 3.8 Impacts avoided by a PV system in inhabitant equivalents.

of normalizing impacts. The values in inhabitant years for the different indicators are given in the table below, which is based on data from IFEN, CITEPA, ANDRA, ADEME and the ministry for industry. Data from the bibliography[271] were used for the damage-orientated indicators.

Table 3.4 Average impacts per inhabitant per year.

Theme	Unit	Inhabitant years
greenhouse effect	kg CO_2	8680
primary energy	kWh	48670
acidification	kg SO_2	62.3
smog	kg C_2H_4	19.7
eutrophication	kg PO_4^{3-}	38.1
water	m^3	339
radioactive waste	dm^3	0.51
other waste	kg equ.	10400
human toxicity	DALY	0.0068
eco-toxicity	$PDF.m^2.yr$	13700

3.1.9 Other life cycle assessment tools for buildings

The company ECOBILAN, whose team is currently part of the PRICEWATER-HOUSECOOPERS group, has developed TEAM BATIMENT[272], which is mainly used to produce environmental and health declarations for materials and components (FDES[273]). This tool could be used to ascertain a building's life cycle assessment with

---

[271] Jolliet Olivier, Saadé Myriam and Crettaz Pierre, Analyse du cycle de vie, comprendre et réaliser un écobilan Presses Polytechniques et Universitaires Romandes, Lausanne, 2005
[272] http://ecobilan.pwc.fr/fr/boite-a-outils/team.jhtml
[273] Cf. la base de données INIES, www.inies.fr

the quantities of each material and the energy consumption, but its complexity makes it less suitable. The CSTB has developed ELODIE[274], which focuses on the construction phase and is based on the FDES. Similar tools are being developed in other countries: LEGEP[275] in Germany, ECOQUANTUM[276] in the Netherlands, ENVEST 2[277] in the United Kingdom, BECOST[278] in Finland, OGIP[279] in Switzerland, ECOEFFECT in Sweden[280], ECOSOFT in Austria[281], ATHENA[282] in Canada and BEES[283] in the USA. The advantage of LEGEP is that it is linked to a software programme that calculates construction costs. EQUER connects the life cycle assessment to the thermal simulation, thus integrating the influence of the choice of products on energy consumption and the environmental impacts linked to this consumption.

A comparative study of the different European tools was made by the thematic network PRESCO (Practical recommendations for Sustainable Construction) coordinated by Belgium's Centre Scientifique et Technique de la Construction. The first case study involved a simple parallelepiped (with concrete sides) heated using electricity, and enabled a comparison between data on concrete and electricity production, and an analysis of the methods: hypotheses on the transportation of materials, recycling, environmental indicators calculated, etc. A second comparison involved a gas-heated Swiss house with a wooden frame. An example of the results relating to greenhouse gas emissions is shown below.

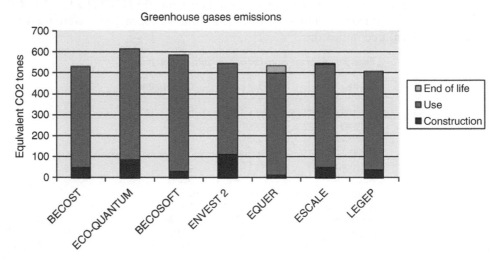

Figure 3.9 Comparison of 7 European Building LCA tools.

[274] cf. http://ese.cstb.fr/elodie/default.aspx
[275] cf. http://www.legep.de/
[276] cf. http://www.ivam.uva.nl/en/
[277] cf. http://envestv2.bre.co.uk/
[278] cf. http://virtual.vtt.fi/virtual/proj6/environ/ohjelmat_e.html
[279] cf. http://www.ogip.ch
[280] cf. www.ecoeffect.se
[281] cf. http://www.ibo.at/de/ecosoft.htm
[282] cf. http://www.athenasmi.ca/
[283] cf. http://www.nist.gov/el/economics/BEESSoftware.cfm

The difference between these tools is $+/-10\%$ over the entire life cycle. Some methods consider a CO_2 balance of nil for wood, with the idea that the CO_2 captured during the wood's growth (photosynthesis) is re-emitted into the atmosphere sooner or later. Other methods consider storage during the manufacturing phase, then different end-of-life emissions depending on the type of treatment (incineration with or without energy recovery, landfill with or without methane recuperation, re-use of wood, etc.). This explains some differences in the construction and end-of-life phases.

Many general tools integrating certain life cycle assessment data have been developed, and they are presented in paragraph 3.7.

3.1.10 Conclusions

The first sensitivity studies showed, for most indicators, the preponderance of flows (energy, water, household waste) in relation to the role of materials that constitute the envelope of contemporary buildings. The only significant contribution concerns waste. However, this sometimes results from the definition of environmental indicators. For example, the resource depletion indicator is defined in relation to the global scarcity of the materials used. Yet little rare matter is included in buildings, for which cheaper materials are preferred. The resource depletion indicator is thus very low, even though in some cases local shortages can occur (e.g. quarry depletion). In addition, the relative contribution made by the materials tends to increase as flows, in particular energy flows, drop. The balance sheet of a passive house, which uses little energy, is presented in chapter 5.

Human toxicity indicators are still imprecisely evaluated, and the location of emissions is not taken into account. There is still a high level of uncertainty regarding data on the manufacture of materials in this theme. This concomitance between the low sensitivity of results and the high imprecision of data and indicators means that it is currently still difficult to use a life cycle assessment to choose building materials. Such an approach will require improving the precision of databases and increasing knowledge of environmental indicators (toxicity, CO_2 balance for wooden materials).

On the other hand, a life cycle assessment for a simplified evaluation of environmental quality, considering the most significant flows of matter and energy, can currently be used on projects in the design stage to study and compare profile improvements.

Optimization based on an ecoprofile analysis provides more general information than the simple energy analysis done up to this point, and may make it possible to synthesize several components of environmental performance (e.g. transportation of people and solar exposure when choosing a building site or urban morphology). A life cycle assessment therefore provides a useful addition to the energy analysis carried out up to now. Initial experiments made by engineering and architecture firms have shown that this method can be used in the building sector. The way it identifies the most influential parameters and uses default values for the other parameters makes it a simplified tool for professionals. Life cycle assessments therefore appear to offer a rigorous, scalable and realistic method for evaluating the quantifiable aspects of environmental quality.

3.2 THERMAL SIMULATION

The sensitivity studies presented above show how important energy aspects are in a building's overall environmental balance. We therefore present below a tool for evaluating heating and cooling loads. To make it easier for designers to implement low-impact solutions, this tool integrates some techniques like thermal and photovoltaic solar systems. The thermal analysis can also be supplemented by a comfort evaluation. We have chosen a simulation method based on a dynamic model of buildings. The models can be "multi-zones", so as to split the building into several areas depending on their function, orientation, etc.

This multi-zone, dynamic simulation method includes techniques for reducing models and object-orientated programming that make it accessible to professionals, architects and engineering firms. A software programme developed along these lines, COMFIE, works with personal computers. The calculation involves reducing a model to finished differences through a modal analysis, with non-linear phenomena or variable parameters being introduced during the simulation phase. Several validation, experimental or inter-model studies have been carried out for different building configurations.

The software programme has been used in numerous studies, e.g. to avoid air conditioning in a tertiary building, assess the performance of bioclimatic houses and solar roofs, and test new technologies like transparent insulation.

3.2.1   Modeling principles

For the thermal analysis of a building, a dynamic tool is needed to model heat storage and evaluate useful solar gains. This is why most detailed models choose simulation. In this family of tools, models with finite differences are the most common because they can take highly varied phenomena into account. Developed on micro-computers, this kind of tool used to be restricted to mono-zone cases (a single air temperature in the whole building). For example, the CASAMO[284] programme.

Model reduction techniques have made it possible to describe buildings in more detail, considering several thermal zones, with a reasonable calculation time (several minutes to simulate a year). A simplified simulation is thus proposed in which a model with finished differences is reduced by modal analysis. This method was initially developed for mechanics before being applied to thermal energy[285]. The theory has been expanded[286] to encompass different applications in space heating[287].

The main phenomena linked to thermal behaviour of buildings can be represented or approached by linear equations. When this kind of linear system can be isolated,

[284]Gabriel Watremez, Dominique Campana, and François Neirac, *Elaboration d'un logiciel sur micro ordinateur pour l'aide à la conception des bâtiments en pays tropicaux secs*, final REXCOOP report, 1985
[285]C. Carter, *A validation of the modal expansion method of modelling heat conduction in passive solar buildings*, Solar Energy, 23, No. 6, 1979
[286]Patrick Bacot, Alain Neveu, and Jean Sicard, *Analyse modale des phénomènes thermiques en régime variable dans le bâtiment*, Revue Générale de Thermique, No. 267, Paris, 1984
[287]J.J. Salgon and A. Neveu, *Application of modal analysis to modelisation of thermal bridges in buildings*, Energy and buildings, October 1987

model analysis can be used. The modelling of linear phenomena is thus simplified. This simplification takes up a fair amount of calculation time since it involves diagonalizing a matrix and changing the base. If the linear system does not vary or of the simulation period can be broken down into a restricted number of intervals over which the system is invariable, then the time investment is worthwhile since the reduction of the model accelerates the simulation.

However, some phenomena are represented by non-linear equations (e.g. natural convection), and the linear system may be variable (closing a shutter increases a window's thermal resistance). These equations are therefore separated from the fixed linear system, and coupled in simulation phase to the reduced modal model. To ensure that the modelling is homogeneous, simplified equations take non-linear phenomena into account.

3.2.2 Main hypotheses and limits of the model

The model is based on the "thermal zone" concept: a sub-set of the building considered at a homogeneous temperature. Although this description is generally a good approximation of reality, the model is not suitable for rooms with very high ceilings in which the air is stratified (e.g. atriums). In the future, it may be possible to split these rooms into several zones (e.g. ground floor and mezzanine) and consider an inter-zone exchange of air with a simple calculation, but this is not currently done.

Fluid mechanic equations are not included in our simplified simulation (although currently under study). Exchanges of air are therefore dealt with using correlations[288]. Air infiltrations are not calculated, since this would involve knowing wind speeds and directions on the same site once the building is constructed, but a method is currently being developed. Indications are provided to evaluate exchanges of air according to the configuration. Equipment is defined only by maximum power, heat and air conditioning, and the yield (heat pumps are currently being modelled). Control is described by the temperature settings (over 52 weeks × 7 days × 24 hours), and by the location of the thermostat. The priority is on the envelope: the objective is to make savings at source on the building's energy requirements. Once the work has been done on the envelope, the development of a data format common to several software programmes means that the project can be transferred to a programme like the equipment-focused TRNSYS (although cumbersome to implement and so less suitable for the early project stages).

Convective and radiative transfers on the wall surfaces are combined into a single exchange coefficient "h". As a consequence, the zone temperature is not air temperature, but is closer to the operative temperature, which is a weighted balance of the air temperature and the temperature of the surfaces. This temperature is a good indicator of thermal comfort, based on the hypothesis that the occupant sets the thermostat to obtain a comfortable temperature: he or she can compensate the cold surface effect by slightly raising the setting. It is therefore the temperature that is controlled and not the air temperature.

The solar radiation entering the zone through the windows is spread over the different opaque walls pro rata to the surfaces, after having deducted, depending on

---

[288] Dominique Blay, *Comportement et performance thermique d'un habitat bioclimatique à serre accolée*, Bâtiment-Energie, No. 45, 1986

the absorption factors, the proportion redirected towards the outside. This radiation is therefore considered to be diffuse, without following the "sun spot". It is difficult to know how spaces will be furnished, and in practice, direct radiation is reflected by furniture. The distribution of flows, taken to be diffuse, is therefore probably a fairly realistic hypothesis.

As we saw in the preceding paragraph, non-linear phenomena or those with variable parameters are uncoupled from the fixed linear system. This can lead to rapid dynamic differences. In practice, the tool is used with a half-hour time step to evaluate heating/cooling loads, and from half an hour to one tenth of an hour to obtain more precise temperature profiles (study of intermittence or summer comfort). The analysis does therefore not include very short time intervals, for which more detailed models would be necessary.

Phase change mechanisms are not included in the current model for humidity or latent heat storage. To consider humidity, enthalpy balances could be established for each zone, similar to those for air exchanges. Integrating phase change material is more problematic, because its thermal coupling with the rest of the envelope does not only concern the zone's temperature. It can in fact be integrated into a wall, which requires putting an additional driving force inside the wall.

The modal reduction method does not therefore appear to be an obstacle to considering the complex phenomena described above, which are problems common to all simulation tools. The advantages of reduced models in terms of less calculation time is that the building can be described in more detail (number of zones) and/or more thorough sensitivity studies can be carried out.

3.2.3 Resolution algorithm

It is fairly standard practice to create a model with finite differences[289], and in this case it is done for each zone. The walls are divided, one-dimensionally, into grids for which a thermal balance is drawn up taking the temperature to be uniform. To make sure that this uniformity hypothesis is not too far from reality would in theory involve dividing up into very small grids. However, the objective is to create a tool on machines accessible to most professionals, which imposes limits on the size of the model. The compromise chosen here involves placing the small number of grids imposed by the computer limitations in such a way that the temperature's uniformity is at a maximum.

The first idea is to avoid grouping into a mesh layers of material separated by an insulating material. Next, the number of meshes must be higher in solid walls than in light partition walls. Lastly, the outputs are the temperatures in the different zones of the building, and these are influenced more by the internal faces of the walls, which are themselves influenced by variations in heating power (controlled equipment, intermittence, etc.) than by the external faces. It is therefore possible to define smaller meshes at the internal face of the wall. A geometric ratio r links the thickness of the successive meshes: if e is the thickness of the most inner mesh, its neighbour has a thickness of r.e, the next one, r^2.e, etc. The value of r can be modified, as can that of the number n of meshes placed in the solid walls.

---

[289] Alain Neveu, *Etude d'un code de calcul d'évolution thermique d'une enveloppe de bâtiment*, Doctoral Thesis, University Paris VI, 1984

The meshes, defined by n and r, do not generally correspond to layers of material. The physical properties of the different materials that constitute a mesh are therefore combined: the masses and thermal resistances of these materials are added together.

We end up with a parameterable mesher (with relation to n and r), which can be used to construct models of different detail. This meshing can be qualified as automatic: the walls are "examined" to determine the number of meshes available so that the user does not need to define these choices.

Different values of n and r were compared[290]. For many types of wall, three meshes appear sufficient to obtain almost identical results (within a tenth of a degree) at a reference corresponding to 20 meshes. The value r has little influence, a value of 3 was also chosen. An internal wall in a zone is divided into analogous meshes.

Insulating material never contains meshes, since its thermal capacity is low compared to other materials. Insulation is therefore simply modelled by thermal resistance, but its thermal capacity is added to that of the two neighbouring meshes.

No meshes are placed in windows: the surface area of windows is high compared to their volume, and the steady state is supposedly rapidly reached in these components. The variable thermal resistance resulting from the use of shading devices (blinds, shutters, etc.) is taken into account in the simulation by introducing a heating power equivalent to the drop in losses.

Air, furniture and any light partition walls in the zone are grouped into a single mesh. The volume of furniture is taken to be small in relation to its exchange surface area, and it is considered to be almost the same temperature as the zone.

During the meshing, the climatic driving forces are listed so that they can later be calculated hourly from the global horizontal and diffuse radiation. Climate is represented by a standard year, e.g. a Test Reference Year[291]. Users can employ different types of data provided they respect a certain format.

Each zone model is then reduced by modal analysis. The first validation studies (cf. below) showed that three modes are generally sufficient to simulate the dynamic behaviour of zones. The first mode, associated with a greater time constant, represents the evolution towards the steady state. The other two represent more rapid dynamics associated for example with daily variations in sunshine, controlled heating power, etc. The use of a software programme in some specific configurations (especially close coupling with the ground) led to an increase in the number of modes: 10 modes are now retained.

The reduced models are then coupled[292]. This involves grouping the matrix subsystems of each zone into a general building model. External driving forces are separated from the interface temperatures, which constitute the coupling variables. The overall system is then integrated according to the time step given by the user (in practice 6 minutes to one hour), and the coupling variables are eliminated.

[290]Bruno Peuportier and Isabelle Blanc Sommereux, *Simulation tool with its expert interface for the thermal design of multizone buildings*, International Journal of Solar Energy, 1988

[291]H. Lund, *Short Reference Years and Test Reference Years for EEC countries*, EEC Contract ESF-029-DK, 1985

[292]Isabelle Blanc Sommereux and Gilles Lefebvre, *Simulation de bâtiment multizone par couplage de modèles modaux réduits*, CVC, No. 5, May 1989

Non-linear and/or variable phenomena are coupled to this global modal model via driving forces for power injected into the different zones. Exchanges by ventilation, for example, are added to the heating power, the internal gains, and the heat yielded by the occupants, to make up a single driving force for each zone.

3.2.4 Computer development

This software is written in Pascal language, and its object-oriented programming facilitates the development of new components: for example photovoltaic systems[293] and ground-coupled heat exchangers[294]. The calculation described above takes less than one minute per zone over a heating season.

To make it easier to use the software, the building is described in the form of a structure of objects linked by pointers. Basic components, materials, glazing, wall coatings, etc. are combined to form more complex structures: walls, zones, entire building. The behaviour of the occupants, linked to the use of the building (accommodation, offices, etc.) is defined in an occupancy scenario containing the profiles of temperature set points, ventilation and internal gains for each day of the week.

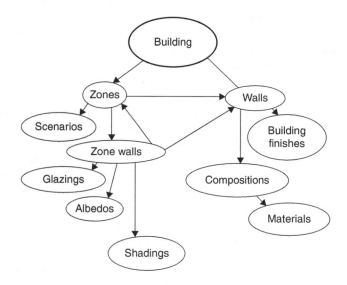

Figure 3.10 Example of a model of a building for thermal analysis.

A pointer links each object to the others: a thermal zone contains pointers on its walls, and each wall contains pointers on windows, shading, etc. A project description

---

[293] Alain Guiavarch, Etude de l'amélioration de la qualité environnementale du bâtiment par intégration de composants solaires, Doctoral thesis, Université de Cergy-Pontoise, November 2003

[294] Stéphane Thiers and Bruno Peuportier, *Modélisation thermique d'un échangeur air-sol pour le rafraîchissement de bâtiments*, Journée thématique SFT-IBPSA-France, Aix-les-Bains, April 2007

can be more or less complex: e.g. a large façade can be broken down into several zone walls to more precisely evaluate the effect of a mask.

The advantage of this kind of structure is that it makes it easier to modify, add, delete or replace an object at any level. This modification is structured: if a material is modified, the modification is automatically propagated to all related wall compositions, e.g. concrete is replaced by brick in the whole building. However, it is also possible to modify only one of the compositions so that all walls with that composition will be modified: e.g. concrete will be replaced by brick in the "walls" composition, but floor slabs will not be modified. Lastly, a single surface can be modified by replacing its composition with another: e.g. concrete replaced by brick only on the southern façade. Thanks to this system, it is very easy to compare design alternatives and therefore refine a design sketch according to the thermal analysis.

3.2.5 Validating calculations

Several validation studies have been done to test the hypotheses of simplified simulation, and in particular model reduction. Different detailed simulation software programmes, e.g. ESP[295], have become references, as have several experimental results:

- comparison with experimental measures on the PASSYS test cell at Stuttgart University[296];
- comparison with ESP in the case of a sunspace by Santi Vitale (ENEL, Italy);
- comparison of annual energy consumption compared to other simplified tools (SUNCODE, APACHE) and sensitivity study of a south-facing glazed area, by John Littler (Polytechnic of Central London, United Kingdom);
- validation of transparent insulation[297].

The figure below shows an example of a house in the Aurore solar housing estate described in chapter 5, where a transparent insulation technique has been applied to the southern façade. The temperatures obtained with COMFIE are compared to the values given by the detailed ESP model. The graphic representation is in the form of a histogram: the y axis shows the number of hours during which the temperature of the house comes within the temperature interval figuring on the x axis.

Another validation tested the correlations between natural convection exchanges in a sunspace built against a test cell at the nuclear studies centre in Cadarache (figure below). The simulation gives slightly higher levels than the measures, especially during cloudless nights. One explanation is the simplified modelling of radiative losses towards the sky: the atmospheric temperature in the model is close to the temperature of the sky, which is lower in reality, especially in the absence of cloud cover.

---

[295] J.A. Clarke, *Energy simulation in building design*, Adam Hilger Ltd, Bristol and Boston, 1985
[296] Bruno Peuportier, *Validation of COMFIE*, Rapport C.E.C., Université de Stuttgart (I.T.W.), 1989
[297] Bernd Polster, *Design of transparently insulated solar buildings*, Diplomarbeit, 1991

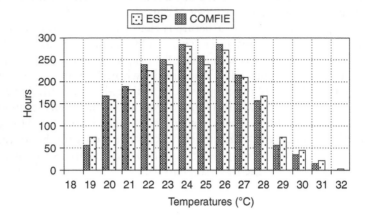

Figure 3.11 Example of a comparative histogram (house with transparent insulation).

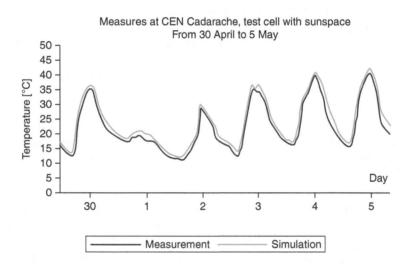

Figure 3.12 Comparison of simulation and measurement (case of a sunspace).

The representation of behaviour dynamics was studied in a comparison with measurements made on an EMPA test cell in Switzerland[298]. The results are shown below.

The simulation underestimates the temperature by around 1°C at some points, but the representation of dynamic behaviour is satisfactory.

A more systematic study is proposed in an ASHRAE standard (American Society of Heating, Refrigerating and Air-Conditioning Engineers). The software studied is compared to 8 other multi-zone software programmes (TRNSYS, DOE, SUNREL,

---

[298] International Energy Agency, task 34

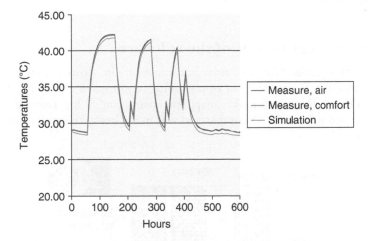

Figure 3.13 Comparison of simulation and measurements (dynamic).

ESP, etc.) on around 30 cases. Each case corresponds to a variation of one parameter (e.g. the orientation of the building, the reflection factor of the walls, etc.). This series of tests can be used to verify the model's sensitivity to the different parameters. The results obtained for COMFIE are shown in the figure below.

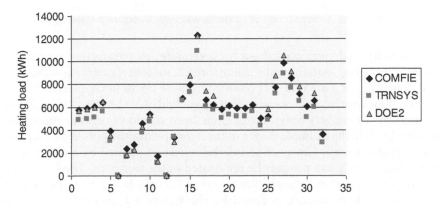

Figure 3.14 Comparison of different simulation tools.

The values obtained by COMFIE are in general fairly close to the maximum interval formed by the results of other software programmes. Overall, the sensitivity closely corresponds to that of the other software programmes except in three cases where the emissivity of the external side of the walls is very low (which is rare in practice). Note that only the ESP result is available for these three cases. The model is therefore judged to be sufficiently reliable given its usage.

The main conclusions of these studies are as follows. Some simplifying hypotheses have little effect on the precision of calculations, e.g. neglecting the termal mass of windows, combining convective and radiative exchanges, considering solar radiation transmitted by the windows as diffuse, or reducing the number of modes in reduced models.

3.2.6 Several applications of the software

The thermal study of the "Aurore" solar housing estate in Mouzon (Architect: Jacques Michel, cf. chapter 5) provided an opportunity to compare different types of solar walls and evaluate the benefits of transparent insulation[299] for Trombe walls[300] (cf. figure below) and solar roofs. A comparison was made between simulations and onsite measurements[301].

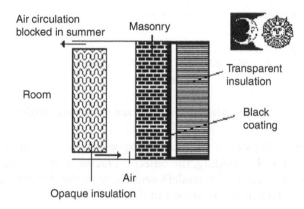

Figure 3.15 Cross-section of a Trombe wall with transparent insulation.

A sensitivity study was carried out for several design parameters: performance of transparent insulating materials, absorbing surface, masonry, orientation and inclusion of solar components. This study is presented in more detail in chapter 5.

Other bioclimatic houses have been studied (Architect: Henri Mouette). If the masonry is relatively heavy, direct gains from windows or solar roofs can halve the heating load. In the example shown in the figure below, the pulsed air that heats the building circulates in the sunspace and in the collectors created by a transparent roof covering. The new air can possibly be pre-heated in the sunspace. The heavy masonry is used to store energy and absorb excess heat so that the house remains comfortable.

A school with an atrium, designed by the architect Jean Bouillot, was studied considering six thermal zones: primary classrooms, nursery classrooms, leisure areas, exercise rooms, atrium, technical areas. The southern façade includes a lot of glazing and is protected by horizontal overhangs above the windows. The central atrium used for circulation supplies additional natural light for the classrooms. The study looked at the distribution of thermal mass and the design of the atrium and reflecting shutters.

---

[299] A. Goetzberger, *Special issue on transparent insulation*, Solar Energy, vol. 49, No. 5, November 1992

[300] cf. http://outilssolaires.com/glossaire/maison-bioclimatique/mur-rombe+a361.html

[301] Bruno Peuportier and Jacques Michel, *Comparative analysis of active and passive solar heating systems with transparent insulation*, Solar Energy, vol. 54, No. 1, January 1995

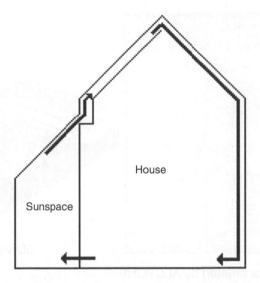

Figure 3.16 Example of air solar heating (sunspace and collectors).

According to calculations, these solar gains compensate for almost 30% of losses (significant air renewal in classrooms).

An International Red Cross Committee building in Geneva (designed by Elio Marcacci) avoids the need for air conditioning by using the coolness of the basement, which is used to store medicine, and adding blinds on the roof. For the same building, a correlation method predicted a temperature of 32°C in the offices, which would have made air conditioning necessary.

In another building (Cité des Sciences et de l'Industrie, Paris), a comprehensive software programme on equipment produced a detailed study of systems, but not of the envelope. A study of the envelope would have shown the considerable gains via the windows, and the air conditioning requirements (higher than for the heating) could have probably been reduced with suitable shading. This problem occurs in numerous tertiary buildings, where the surface area and the orientation of glazed areas have not been studied from a thermal point of view. The F. Mitterand library in Paris is another good example: wooden shutters had to be added on to protect books from the sun.

The thermal consequences of architectural choices are often overlooked due to methods that are too simplistic or focused on equipment. The result is waste, even when equipment is well managed. The emphasis should be on encouraging a better balance between aesthetic and functional aspects right from the architectural plans, and software can no doubt play a part in this area.

3.2.7 User interface

Inputting data and formatting results in a study report take time, and these tasks must be made easy to fit in with the obligatory time constraints of professional

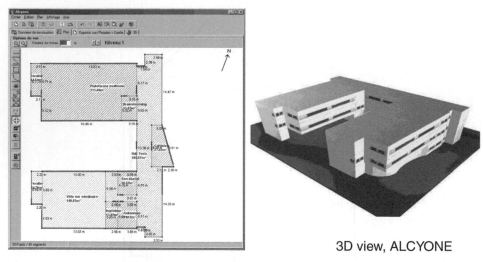

2D plan imported or created by ALCYONE

3D view, ALCYONE

Figure 3.17 ALCYONE software, 2D-3D graphic input.

practice. In addition, graphic utilities can be useful to visualize some parameters (e.g. solar radiation and mask effects) or results (temperature profiles). To this end, the COMFIE programme presented above has been supplemented by a graphic input tool ALCYONE and an interface called PLEIADES[302], illustrated by several examples below.

Study of a site

Neighbouring buildings, trees and hills can create masks that block direct sunlight at some times of day and seasons. A diagram like the one below can be used to evaluate these mask effects.

This type of representation can for example help evaluate the possibility of installing solar collectors at the location considered: in the above case, an 18-storey tower creates a mask to the south-east, but the diagram shows that the shading effect lasts for less than two hours a day. It is therefore possible to install a solar water heating system.

Study of solar protection

The example chosen here concerns horizontal protection (overhang) located above a window. The aim is to reduce solar radiation in the summer, but as little as possible during the heating season. The diagram below can be used to evaluate the reduction in

---

[302]cf. http://www.izuba.fr

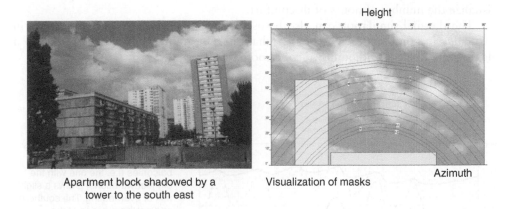

Height

Azimuth

Apartment block shadowed by a tower to the south east

Visualization of masks

Figure 3.18 Evaluation of the solar exposure of a façade (PLEIADES software).

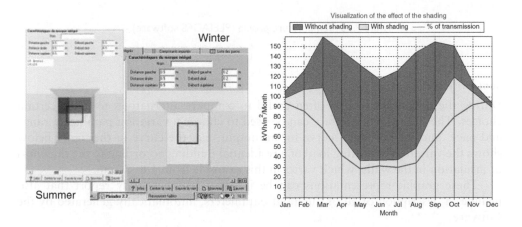

Figure 3.19 Evaluation of efficiency of solar protection (PLEIADES software).

radiation (difference between the dark curve, incident radiation, and the light curve, transmitted radiation) for each month.

On the southern façade, sunlight from a clear sky (dark curve) is minimal in the summer (the sun is high and so the rays are very oblique in relation to the façade) and fairly considerable in winter, when it is useful; the overhang is very efficient in reducing radiation in summer without penalizing it in the winter. A sensitivity study on the length of the overhang can be used to size this parameter.

Temperature profiles and histograms

Temperature profiles can be used to verify that the thermal comfort level is reached, e.g. for offices where the thermostat set point is lower at night. Temperature histograms give a concise representation of temperatures over a season and make it possible to visualize the number of hours of discomfort.

The thermal simulation can also be used to evaluate the risk of overheating, in particular in zones exposed to solar radiation. The figure below corresponds to a glass-fronted balcony adjoining a living room.

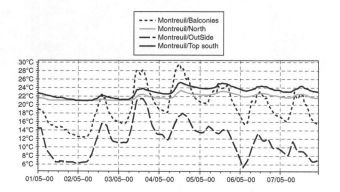

In this example the temperature curve for the glass-fronted balconies (i.e. the one with the biggest variations) shows a slight risk of overheating. The southern and northern zones of the building (cf. the two other curves) remain comfortable during the simulation period considered.

Figure 3.20 Temperature profile (PLEIADES software).

Sensitivity studies

During the design phase, it is often useful to compare alternatives to make certain choices. The PLEIADES interface can be used to choose a variation parameter, range and pace, and then obtain a curve for an output: e.g. the curve in the figure below shows the variation in heating load of a residential building according to the thickness of insulation in the walls, which varies in this example from 1 to 20 cm.

These graphic representations can be integrated into a study report that also includes a description of all of the input parameters, automatically generated by the software.

This type of interface makes the simulation accessible to professionals, since a building study takes a few hours, which corresponds to the time allocated to this type of evaluation in a project's budget.

3.3   LIGHTING CALCULATIONS

Consumption due to lighting constitutes a significant proportion of overall energy requirements in most buildings, in particular non-residential premises like offices, in which heating loads are relatively low due to higher internal gains (i.e. people, computer equipment). Overheating is often a problem, and artificial lighting often constitutes a significant share of the cooling load.

Bioclimatic design aims to reduce energy demand for heating and lighting by recuperating solar gains, while minimizing overheating problems caused by excessive penetration of solar rays. A simulation model that deals with both solar heating and

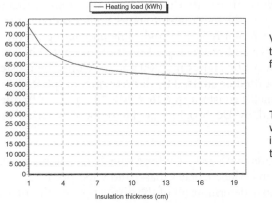

Variation in heating load according to the thickness of insulation in the façades.

The heating load drops sharply with the first few centimetres of insulation, but the influence of the thickness diminishes afterwards.

Figure 3.21 Sensitivity analysis (PLEIADES software).

the thermal consequences of lighting can be used to study the benefits of passive solar heating and lighting, and to optimize the size and orientation of windows.

This paragraph presents a coupling between the COMFIE software described above and a lighting model that can be used to evaluate the level of daylighting and energy consumption due to artificial lighting. COMFIE was developed as a design aid that is both easy to use and sufficiently sensitive to design parameters. The aim was therefore to find a compromise between simplicity and precision, adapted to professionals' needs; the same approach is adopted here for lighting calculations. A simplified, hour-by-hour simulation of the lighting in a building is set up. The thermal behaviour of buildings is analyzed after breaking them up into zones. These zones group rooms that are taken to behave homogeneously, and are defined by their surface area, orientation, the composition of the surfaces that create boundaries, and by the size (length and width) of window surfaces present on each surface area. The same structure can be used as for the objects defining the building, although the lighting analysis generally requires breaking the zone down into rooms (except when all rooms and their environment are identical).

The calculation includes the following stages:

- establish the incident solar light flux;
- model the transmission of this flux in the zone;
- calculate the natural internal lighting on a reasonable number of fractions of the work surface studied in the room;
- usage of artificial lighting;
- visual comfort (daylighting of areas) and energy consumption for lighting.

3.3.1 Light

The nature of light

At the start of the last century, the notion of "wave-particle duality" emerged in the theory of light: light was initially envisaged as a wave transmitted through an atmosphere by sinusoidal movement. This led to Maxwell's theory (1872), in which light is described as a ray of energy propagated in the form of an electromagnetic wave. The particle theory is based on the fact that luminous bodies radiate energy in the form of particles, emitted intermittently in a straight line and acting on the eye's retina by stimulating the optic nerve and producing a sensation of light. The modern version of this theory was put forward by Planck in 1900 and developed by Einstein in 1906. It affirms that energy is absorbed or emitted by discrete quanta (photons), with a magnitude of hv, where h is Planck's constant (6.626×10^{-34} J), and v a frequency.

The unified theory established by De Broglie and Heisenberg is based on the following assumptions:

- each moving particle is associated with a wave whose length is given by: $\lambda = h/mv$, where h is Planck's constant, m is the particle's mass, and v its speed;
- it is impossible to simultaneously determine all of the distinct properties of a wave or a particle.

Whether light is seen to be in wave or photon form, it is radiation[303] produced by excited electrons in incandescent solids[304], which liberate energy when they return to a more stable state.

The light spectrum

White light, such as sunlight, is the addition of monochromatic rays ranging from violet to red, according to their wavelength.

The limits of the visible spectrum vary slightly depending on observation and the intensity of the beam observed. Conventionally, infrared rays (termed as near, mid or far from the visible spectrum) range from the wavelengths 0.76 μm to 0.1 mm[305]. The values 0.38 μm and 1 nm are usually considered to be the limits of the ultraviolet range. The solar spectrum practically stops at around 0.2 μm, since rays with shorter wavelengths are absorbed by the atmosphere. Visible rays (from 0.38 to 0.76 μm), at equal energy, have very variable light power. The eye's sensitivity to light rays varies depending on their length. The notion of luminous flux implies two conventions:

- a light source transforming a power of 1 W without loss into a ray of length $\lambda = 555$ nm, a yellow-green ray to which the eye is the most sensitive, emits 683 lumen (value chosen to correspond to a former system based on the candela);
- when the wavelength is different, for each value of λ a factor $V(\lambda)$, which varies from 0 to 1, is defined as the luminous spectral efficiency[306].

Bright objects like the moon, clouds and the daytime sky, which only diffuse received light, constitute "secondary" light sources (reflected or transmitted). This diffusion is done without changing frequency except in exceptional cases.

---

[303] Liberation of energy by a source. Energy from a point source is inversely proportional to the square of the distance from the radiation source in the absence of adsorption
[304] Incandescence = emission of light from a hot body as a result of its temperature
[305] 1 μm = 10^{-6}m (micron), 1 nm = 10^{-9}m (nanometre)
[306] Marc La Toison, *Matériels et projets*, Techniques de l'ingénieur, C 3341

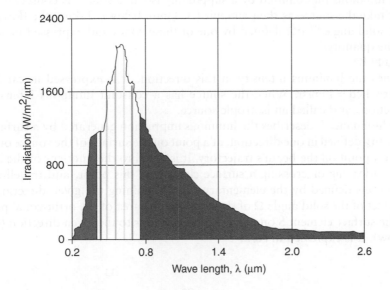

Figure 3.22 Spectral distribution of sunlight.

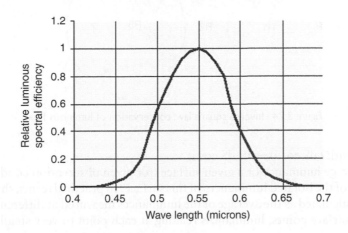

Figure 3.23 Sensitivity of the eye according to stimulus wavelength.

Definitions

- The quantity of light emitted, by unit of time and by a source of light, is called the luminous flux Φ, measured in lumen (lm). A lumen is therefore defined in terms of radiated power weighted by the sensitivity of the human eye.
- The luminous flux received by a surface with an area dS defines the lighting E of the surface at the point considered:
- E = dΦ/dS
- Light is expressed in lumen per square metre, or lux (lx). It is an additive quantity.

- The luminous flux emitted by a supposedly isolated source is conserved in cones that take the source as their summit (cf. figure below). Taking the flows emitted in a solid angle[307] $d\Omega$ defined by one of these cones and expressed in steradians sr, the quantity:
- $I = d\Phi/d\Omega$
- defines the luminous intensity in this direction. It is expressed in candela (cd), where $1\,cd = 1\,lm/sr$. When the source has a constant luminous intensity in all directions, it is called an isotropic source.
- The luminance L describes the luminous impression generated by a surface. It is a quantity defined in one direction, at a point on the surface of the source or receiver, or at a point on the beam's trajectory. It is equal to the quotient of the luminous flux Φ leaving or crossing a surface element at this point, and travelling in the directions defined by the elementary cone containing the given direction, by the product of the solid angle Ω of the cone and the area of the orthogonal projection of the surface element S on a plane perpendicular to the given direction (cf. figure below). It is expressed in cd/m^2.

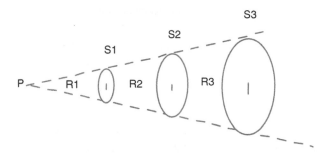

Figure 3.24 Inverse square law: conservation of luminous flux.

- $L = d^2\Phi/(d\Omega\ dS\ \cos i) = dI/(dS\ \cos i)$
- The average luminance of a given surface (for given observation conditions) is the average of the isolated luminances of this surface. In practical terms, this luminance can be calculated as the average of the luminances measured at different judiciously chosen surface points, luminances relating at each point to very small, equal solid angles.
- The working plane (or work surface) is the reference surface considered, constituted by a plane on which an activity is carried out that requires a certain amount of lighting. For interior lighting, and unless otherwise indicated, the plane is conventionally a horizontal plane located 0.85 m from the ground and demarcated by the walls of the premises.
- Taking systems' energy consumption into consideration lastly involves defining the luminous efficiency of a source of light. For electromagnetic radiation, this is

---

[307] A solid angle is a measurement in space of radiuses leaving a point in a euclidean three-dimensional space. It is equal to the area intercepted by all of the chosen radiuses on the surface of a unit sphere centred at the source point. The international unit is a steradian.

defined by the quotient of the luminous flux over the energy flux of the radiation. For an electric source, it is the quotient of the luminous flux produced over the electric power consumed. In both cases, it is expressed in lumen per Watt (lm/W).

3.3.2 External illuminance

Sources of received flux

On Earth, we receive two types of luminous radiation from the atmosphere: direct radiation from the sun, and diffuse radiation from the sky. The sun is the primary source of natural light, putting aside the light coming from the other stars. Light from the sky is the same light that has been transmitted, diffracted, and diffused by the atmosphere with its molecules and particles in suspension.

On a building, a third source of light should be considered: the ground, plants and neighbouring buildings that reflect light. These same items, along with some other objects, can also constitute masks that diminish the quantity of light received.

A surface with a given slope and orientation therefore receives light from three origins, in proportions that depend on the sun's position in the sky, the climate conditions, and the characteristics of the site. The light received each hour by the glazed surfaces can be used to determine the light transmitted towards the inside zone considered.

Variations in solar radiation

The Earth's elliptical orbit around the sun modifies the Earth-sun distance throughout the year, leading to a variation of around 3% in the radiative flux received at the edge of the atmosphere. In addition, the Earth turns on itself by 15° per hour, around an axis at an angle of 23.45° in relation to the ellipse. A point at a given latitude and longitude will therefore receive solar radiation that has travelled a variable distance and crossed a variable thickness of the atmosphere (cf. figure below).

During its passage through the atmosphere, radiation is diffused by air molecules, water vapour and dust, and absorbed by ozone molecules, water and carbon dioxide. Since these phenomena are selective depending on the wavelength, the spectral division of solar radiation is modified according to the air mass crossed, and its composition (climate, pollution)[308]. Its luminous efficiency is thus modified, which leads to a change in the distribution of luminance in the sky[309].

Calculating solar driving forces using COMFIE software

The available climate files give, among other things and for a given station (latitude, longitude, altitude), hourly values for horizontal global radiation, horizontal diffuse radiation and normal direct radiation in J/cm^2. The driving forces corresponding to these solar fluxes on the external surfaces of a building are calculated hourly according

---

[308] Duffie, Beckman, *Solar Engineering of Thermal Processes*, Solar Energy Laboratory, University of Wisconsin, Madison, 1980

[309] CSTB, Michel Perraudeau, *Distribution de la luminance du ciel*, Etudes et Recherches, Centre Scientifique et Technique du Bâtiment, Cahier 2305, livraison 295, December 1988

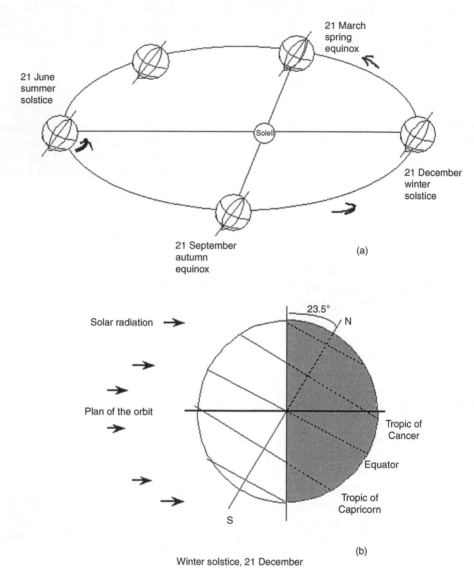

21 March
spring
equinox

21 June
summer
solstice

Soleil

21 December
winter
solstice

21 September
autumn
equinox

(a)

23.5°   N

Solar radiation

Plan of the orbit

Tropic of
Cancer

Equator

Tropic of
Capricorn

S

(b)

Winter solstice, 21 December

Figure 3.25 Geometry of Earth's orbit and tilt of the polar axis[310]. (a) Complete orbit and (b) Enlarged
    details, solstices.

to the orientation and slope of the surfaces using the Liu and Jordan method[311]. Rays
that are diffused and reflected by the ground are considered as isotropes, but it is
indispensable to separate them in the daylighting model. The reason is that these two

[310]Rabl Ari and Kreider Jian, *Heating and Cooling of Buildings*, McGraw Hill, 1994
[311]Duffie, Beckman, *Solar Engineering of Thermal Processes*, Solar Energy Laboratory,
University of Wisconsin, Madison, 1980

fluxes do not have the same incidence direction, and are reflected differently inside the premises.

Radiation is reduced in the presence of masks close by (e.g. balconies, overhanging roofs) or far away (e.g. other buildings, trees). This reduction is characterized by a shadow factor, in relation to close and distant masks and to direct or diffuse radiation. This factor is calculated for each hour[312].

The radiation transmitted (directly from the sun, diffuse from the sky or the ground) can be used to determine the incident light on the glazed surfaces of the building considered. The contribution of light from reflections of the solar flux on the building's external environment other than the ground is currently neglected. This approximation can be far from minimal[313]: a highly reflective south-facing façade may sometimes shine brighter than natural lighting (sun and sky) on an opposite north-facing façade. However, to take this phenomenon into account would require a description of the buildings directly opposite, which would complicate the data input. It would nevertheless be possible to account for reflection on distant masks that are already modelled in COMFIE.

Modelling incident luminous flux

A measurement campaign carried out in Nantes enabled the CSTB to classify five types of sky[314]. This classification is based on a single index: nebulosity I_N:

$$I_N = (1 - CR_M)/(1 - CR_T)$$

where CR, i.e. Cloud Ratio, the index used by the NBS[315], is the quotient of diffuse horizontal radiation on global horizontal radiation, CR_M is the measured value of CR, and CR_T the theoretical value of CR in a clear sky.

The values of I_N associated with each of the five types of sky are given in the table below (column 1). For hourly simulations, the solar radiation (watts) and diurnal natural light (lumens) can be taken to be proportional for each time step[316]. However, to refine the model in relation to constant luminous efficacy, the nebulosity index can easily be calculated and luminous efficacy can be deduced (column 3).

Since the rays are measured in kiloJoule per square centimetre during one hour, the corresponding light is deduced by the factor ($e * 10^3/0.36$), taking an average luminous efficacy of e lm/W. The external light values can thus be deduced from the values, already calculated for heat, of the hourly radiation incident on the glazed surfaces (which takes into account dimming caused by masks and the opaque parts of the glazed surfaces).

---

[312]Peuportier Bruno and Blanc-Sommereux Isabelle, *Comfie, Passive Solar Design Tool for Multizone Buildings, Manuel des Utilisateurs, Version 3.3*, August 1994

[313]Fontoynont Marc R., *Prise en compte du rayonnement solaire dans l'éclairage naturel des locaux: méthodes et perspectives*, Doctoral thesis, Ecole Nationale des Mines de Paris, 1987

[314]Michel Perraudeau, *Distribution de la luminance du ciel*, Etudes et Recherches, Centre Scientifique et Technique du Bâtiment, Cahier 2305, livraison 295, December 1988

[315]National Bureau of Standards, USA

[316]Rabl Ari and Kreider Jian, *Heating and Cooling of Buildings*, McGraw Hill, 1994

Table 3.5 Nebulosity index and luminous efficacy[317].

Type of sky	Nebulosity index	Luminous efficacy
Overcast	$0.00 < IN < 0.05$	100 lm/W
Intermediate overcast	$0.05 < IN < 0.20$	110 lm/W
Intermediate median	$0.20 < IN < 0.70$	120 lm/W
Intermediate blue	$0.70 < IN < 0.90$	130 lm/W
Blue	$0.90 < IN < 1.00$	140 lm/W

3.3.3   Modeling daylighting using the lumen method

The lighting at any point of the premises considered is caused by direct light coming from the outside (directly from the sun and sky, or indirectly after reflection on the ground or neighbouring buildings), and to the internal reflection of this luminous flux on the different surfaces that make up the premises.

The LUMEN method was initially chosen because it works well with the COMFIE programme due to the compatibility of input data. This method is based on the notion of utilance, which corresponds to the efficacy with which light reaches the work surface taking into account reflections on the walls.

The main hypotheses are as follows:

- The shape of the room is necessarily a parallelepiped;
- The solar radiation reflected by the ground and all other reflecting surfaces is taken to be perfectly diffuse and isotropic.
- The use of a more detailed model, by ray tracing, has been implemented by chaining other software.

Luminous flux transmitted by glazing

Direct radiation transmitted by transparent surfaces is supposedly diffused upon contact with the walls on which it falls, and its participation in lighting the premises is jointly evaluated with that of diffuse lighting[318]. The shading scenarios defined for heat can thus be used, as long as the luminous transmission factors are considered.

The transmission factor of glazed surfaces varies depending on the angle of solar rays. It is taken to be similar for all wavelengths, following the same law as in the thermal model. Rectifying coefficients can be used depending on how clean the window

[317]Most bibliographic references take luminous efficacy as from 90 to 140 lm/W for daylight. However, Fontoynont (*Potential for increased efficiency in indoor lighting, Efficient Use of Electricity in Buildings*, IEA: Energy Conservation in Buildings and Community Systems, Workshop Report, Sophia-Antipolis, May 1994) gives a value of 235 lm/W for the efficacy of daylight. More studies therefore need to be done.

[318]In practice, occupants often use curtains and blinds to reduce glare, and this simple model is possibly close to real-life conditions. This hypothesis involves reducing contrasts between the different parts of the room, and thus increases the significance of internal reflections in the evaluation of interior lighting levels. It therefore transforms the lighting obtained by transparent bay windows into lighting obtained by translucent bay windows, or those equipped with net curtains.

is (cf. table below) and the possible presence of a light shaft (cf. figure below). In existing buildings, any deterioration of interior surfaces should be taken into account in the reflection coefficient data.

Table 3.6 Correction factor relating to dustiness of windows.

	Outside conditions	Inside conditions		Window angle in relation to horizontal		
Location	Opacity of window's outside surface	Nature of activity in the building	Opacity of window's inside surface	0 to 30°	45 to 60°	75 to 90°
Rural region or suburban zone	low	clean	low	0.80	0.85	0.90
		dirty	high	0.55	0.60	0.70
Urban or residential zone	average	clean	low	0.70	0.75	0.80
		dirty	high	0.40	0.50	0.60
Industrial zone	high	clean	low	0.55	0.60	0.70
		dirty	high	0.25	0.35	0.50

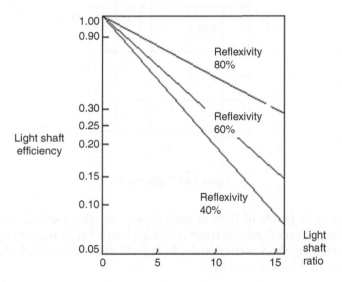

Figure 3.26 Efficiency of a light shaft.

Figure 3.26 is taken from IES[319]. The light shaft ratio is the quotient $(5 \times \text{height} \times (\text{width} + \text{length}))/(\text{length} \times \text{width})$ in relation to the dimensions of the light shaft, and the effiency of the shaft is the proportion of light transmitted.

---

[319]IES, *Recommended Practice for the Lumen Method of Daylight Calculations*. IES Report RP-23, Illuminating Engineering Society of North America, 1989

Evaluating the daylighting of a premises

a)   Dividing up the room to calculate the lighting

In the energy analysis, rooms with the same thermal behaviour are grouped into the same zone. If rooms differ in their geometry or external environment (e.g. presence of masks), it is preferable to undertake a sensitivity analysis to determine whether lighting results are very different when the studied zone is divided into rooms.

In the model, light coming from a façade decreases with distance in a direction perpendicular to the window, but the decrease on the sides of the window is not considered. Overhead light is taken to be constant throughout the room. If the zone contains different window openings, the lighting calculation will be done window by window, and then added together.

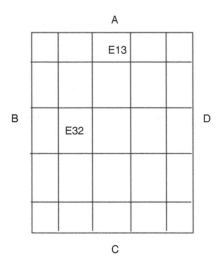

Figure 3.27   Light matrix.

Considering X bands of light in each dimension (this method is only used for parallelepiped premises), the room is divided into X^2 rectangles of light (cf. figure above). Naming the 4 vertical walls A, B, C, D, the resulting light in one of these rectangles (i, j) is expressed by:

$$E_{ij} = e_i(A) + e_j(B) + e_{X-i+1}(C) + e_{X-j+1}(D) + e(Z),$$

where $e_i(A)$ is the lighting at depth i due to the glazed surfaces of the wall and A and e(Z) is the lighting due to the overhead window openings.

b)   Utilance tables

The utilance coefficients[320] of the rooms are generally given at the percentages 10, 30, 50, 70 and 90 of the depth of the room studied in relation to the vertical

---

[320] IES, Recommended Practice for the Lumen Method of Daylight Calculations, IES Report RP-23, Illuminating Engineering Society of North America, 1989

plane including the window. It appears sufficient to only retain 10, 50 and 90, which generally represent the maximum, average and minimum lighting points for the window considered[321].

- The index of the premises K, characterizing the geometry of the part of the premises between the useful plane and the light fixture plane, is used to determine the utilance:
- $K = ab/h * (a + b)$
- where a and b are the dimensions of the sides of the premises and h is the difference in level between the useful plane and the light fixture plane.

For façade lighting, utilance values are supplied by tables in relation to the premises index and the coefficients of the surface reflection (i.e. ground, wall and ceiling). For overhead lighting, the values also depend on the type of opening (plane, dome, clerestories, etc.).

3.3.4 Artificial lighting

As for heat, the equipment has been modelled simply in COMFIE, to avoid requiring detailed input at the architectural design stage. A simplified calculation is therefore also carried out for artificial lighting.

Definitions

- Light fixture: apparatus used to spread, filter or transform light from one or more lamps and comprising, apart from the lamps themselves, all the parts needed to fix and protect the lamps and possibly auxiliary circuits as well as devices for connecting up to the electricity supply;
- Classification of symmetrical light fixtures for interior lighting: according to standard UTC C 71 121[322], any symmetrical light fixture (rotational luminous flux distribution) is represented by the sum of a direct light fixture and an indirect light fixture[323]. Classification is defined according to the luminous flux distribution in the different cones located around the rotation axis, and comprises ten classes of direct light (from A, very direct, to J, very diffuse) and an indirect light fixture class, T;
- Light fixtures are therefore described thanks to their upper yield η_s (yield for the part of the ray emitted upwards) and their lower yield η_i, and are noted as $\eta_s T + \eta_i (A \ldots J)$;
- Ballast: device interposed between the current and one or several discharge lamps that are mainly used to limit the current of the lamp(s) to a required value;

---

[321] The lighting values calculated should respectively be taken as the average values of the surfaces of the room ranging from 0% to 30% of the depth, 30% and 70%, and 70% and 100%.

[322] http://www.afnor.org/profils/activite/electrotechnologies

[323] The CIE (Commission Internationale de l'Éclairage) created a classification of light fixtures by type: direct, semi-direct, general diffuse, direct-indirect, semi-indirect, indirect (IES, 1987). This corresponds to the direction in which the light is transmitted, with direct light fixtures distributing all of their light downwards, and indirect light fixtures distributing upwards.

- Nominal power: power of the lamp indicated in the manufacturer's catalogue. For discharge lamps, this power corresponds to the lamp only, without the consumption of ballast; it therefore needs to be added on to obtain total consumption;
- Depreciation factor, d: relationship between average lighting on the useful plane after a certain duration of lighting usage and the average lighting for the installation considered as new;
- The colour rendering index (CRI)[324]: The CRI gives an indication of a source's capacity to render colours in comparison to a reference source. Conventionally, the CRI ranges from 0 to 100. The reference sources have a colour rendering index of 100 and are the blackbody up to 5000 K and normalized natural light beyond this colour temperature;
- Colour temperature: a source's colour temperature gives an indication of the coloured appearance of the light emitted by the source. Simultaneous use of sources with different colour temperatures is not recommended since it causes visual disturbance, in particular due to the chromatic adaptation of the eye. A very high colour temperature e.g. ($T_c = 6000$ K) corresponds to the blue zone and the overall impression in the premises will be cold. A low colour temperature (e.g. ($T_c = 2800$ K) corresponds to the red zone and the impression will be warm;
- Comment: the colour temperature T_c of a source cannot be rigorously defined if that source has exactly the same chromaticity as a blackbody at T_c K. In other cases, only the "proximal" colour temperature can be defined (i.e. the temperature of a blackbody whose chromaticity is closest to that of the source considered).

Material

Standardization organizations have established security standards for lights: mechanical, electric, thermal and optical security for lighting products. In France these standards are generally delivered and managed by the UTE (Union Technique de l'Electricité), and certified by AFNOR (Association Française de Normalisation). ISO (International Standard Organisation) has delegated electro-technical and electronic standards to the International Electrotechnical Commission (IEC) headquartered in Geneva.

The luminous efficacy of lamps varies from around 10 lm/W for incandescent lights, 20 for halogen, from 65 to 105 for fluorescent lamps, from 40 to 80 for compact-fluorescent lights, and from 30 to 100 for LEDs[325].

Only some of the light produced by a lamp actually reaches the work surface, the rest being absorbed by nearby objects and the fixture, which also influences the shape of the beam it gives out. Like lamps, light fixtures are subject to numerous security and visual comfort regulations. The IEC (International Electrotechnical Commission, publications No. 40 and 52[326]) has established detailed classifications for light fixtures depending on the distribution of their luminous flux in the room, which can be used

---

[324]The DOE reference (1993) includes a table that gives the colour characteristics of lamps (yield and temperature index, see further on)

[325]La Maison écologique No. 46, Aug.–Sept. 2008, www.lamaisonecologique.com

[326]Available from the Association Française de l'Eclairage, Editions Lux, www.afe-eclairage.com.fr

to calculate the lighting values at different points of the room. The UTE C 71 121 standard provides a simplified version of this method, and can be used to determine average lighting on the useful plane and surfaces.

This method was used by the CSTB in the "Mixte" software. Although simplified, it takes into account the exact position of light fixtures, taken to be regularly spaced on rectangular meshes, and only makes sense if the lighting is fairly uniform. It can be used to size an installation (i.e. number, type of power of lights and fixtures) based on the characteristics of the premises and the level of lighting sought on the work surface.

Lighting level per usage

Lighting demand varies widely depending on the buildings considered and the usage of rooms inside these buildings. The choice of appropriate lights and light fixtures depends on the power required, the dispersion of the light beam (light dispersed homogeneously or light source), how often it is turned on, the quality of the light (CRI and colour temperature for atmosphere and aesthetic reasons), and the overall cost (i.e. cost of purchase and luminous efficacy).

Lighting levels are recommended in AFNOR standard X 35-103 (1990). The influence of lighting on visual performance is experimentally characterized by the performances of normal observers aged from 20 to 30, in reference conditions using a diffuse light coming from an environment with uniform luminance. It is thus possible to define a background luminance level for each task, adapted to perceiving details, but in the reference conditions described above. More simply, an average lighting in service is defined to support the required luminance task. This value corresponds to the average lighting in service measured in the middle of the period from start-up to first maintenance[327].

The following paragraph gives, for each type of building, any relevant regulations, the required lighting levels, the atmosphere and the appropriate type of lamp[328].

a) Accommodation
Lighting in accommodation is characterized by the high number of light sources. The rendering of colours is important. The main requirements are aesthetics, light quantity, maximum flexibility, and low cost. The most used sources are therefore: halogen lamps, lamps with built-in reflectors and compact-fluorescent lights.

b) Industrial buildings
In general, the target is low price and consumption, average light quality, and mostly direct lighting, usually resulting in:

- Low-level factories: fluorescent tube lighting with the best possible efficiency levels, except if colour rendition is very important;

---

[327]Lighting designers can calculate initial lighting by applying to the lighting values in paragraph 3.3.6 a depreciation compensating factor corresponding to half of the reduction in flux anticipated at the date of the first maintenance. Other conditions, like the age of observers, the duration and frequency of the observation, and the seriousness of the consequences of installation sizing errors can lead to these values being corrected.

[328]An extremely detailed table of lighting levels can be found in the document (IES, 1987), on pages 2-5 to 2-20. They are classed by generic type, from A to I. Weighting factors are proposed according to age of user, importance and duration of the task, etc.

Table 3.7 Lighting requirements for accommodation (AFNOR X standard 35–103)[329].

Entrance hall	150 lx
Background lighting	200 lx
Bedroom, localized lighting	300 lx
Cooking preparation	300 lx
Office	300 lx
DIY area	300 lx

Table 3.8 Lighting requirements for industry (AFNOR standard X 35–103).

Minimum for visual tasks	200 lx
Heavy machining, diverse industrial tasks	300 lx
Industrial precision work	500 lx
Delicate work	1000 lx
Very delicate work	2000 lx
Animal buildings	50 lx

- Very high factories: high-intensity discharge lamps;
- Supplementary lighting: inbuilt reflector lamps.

c) Offices
    Technical drafting offices: AFNOR standard X 35-103   1000 lx
    Contrast rendition is crucial. Lines of fluorescent lights are usually placed perpendicular to the lines of tables.
    Other offices: some documents recommend 500 lx, but the tendency is to reduce base lighting to 300 lx and add supplementary lighting, e.g. desk lights equipped with 2W LED (light-emitting diode).

d) Rooms with projection screens:
    Screens with light surface   <300 lx
    Screens with dark surface   <1000 lx
    Contrast rendition is very important. Visibility must be good with no reflection. The most suitable light fixtures are low luminance lamps, and fluorescent lamps (tube/compact).

e) Commercial lighting
    The usual requirements are good light quality (Colour Rendition Index – CRI – and high quality), and significant flexibility (inside, shop windows, outside). Solutions include spotlights, standard or low-voltage halogen lamps, and fluorescent lamps (often 58 W tubes measuring 1 m 50, or compact).

---

[329] AFNOR standard X35-103, *Ergonomie – Principes d'ergonomie visuelle applicables à l'éclairage des lieux de travail*, October 1990

Table 3.9 Lighting requirements for commercial premises[330].

Small shops	300 lx
Self-service stores	500 lx
Department stores	1000 lx

f)   School lighting
The standard choice is louvered fluorescent lamps in lines perpendicular to the tables, and special fluorescent lamps for the blackboards.

Table 3.10 Lighting requirements for schools[331].

	Recomm. E, lx	Min. E, lx
Pupil desks	300	150
Teacher desks	400	200
Blackboards	500	300
Drawing tables	700	400

g)   Sports facility lighting
The most common choices are fluorescent lamps and low-luminance lights.

Table 3.11 Lighting requirements for sports facilities[332].

Multi-sports	200 lx
Tennis courts	500 lx
Swimming pools	150 lx

h)   Hotels
Lighting is generally localized. The colour of lamps is important, in particular in bathrooms, and increasingly so with the quality of the hotel. Halogen and fluorescent lamps are usually used.

i)   Hospitals: AFE recommendations and approval by the Ministry for Health.
In bedrooms, lighting is indirect on the ceiling, and very localized on the bed. Stronger lighting is required at the bed head. Fluorescent lamps are widely used. Requirements are very strict for operating theatres, and are stipulated in the technical specifications provided to designers.

j)   Museums: Much more detailed modelling is necessary. Care should be taken to avoid rays that are likely to damage works, and optimize lighting quality. Specialized tools are required.

---

[330] AFNOR standard X 35-103; AFE commercial lighting brochure, regulatory and contractual texts

[331] Ministerial decree, JO 30 March 1965, minimum and recommended lighting

[332] AFE recommendations officialised by the Ministry for Youth and Sport (twice the values for competition training)

Table 3.12 Lighting requirements for hotels[333].

Dining rooms	200 lx
Foyers	300 lx
Corridors and open spaces	150 lx
Kitchens	300 lx

Modelling artificial lighting

a)   Defining an occupancy scenario

For the COMFIE heat calculations, the user defines weekly occupancy scenarios, for which he or she provides the expected values of inside temperature depending on the day and time. The same principle is applied to lighting, this time using lighting values. However, there is a difference due to the fact that lighting must be defined using a two-level approach: atmospheric lighting/supplementary lighting.

Table 3.13 Example of a lighting scenario for a living room[334].

Day	Time	Uniform lighting (lx)	Supp. lighting (lx)
Monday	19	0	300
	20	200	0
	21	0	300
	22	0	0

The next step is to define the criteria for turning on lights. In general, lights are turned on when the daylight level drops below 80% of the threshold value indicated in the occupancy scenario. In line with the current possibilities of software to calculate artificial lighting, three criteria have been selected:

- The zone's average lighting is taken as the threshold value;
- The weakest lighting in the zone is taken as the threshold value (extreme importance of light for the task done, or "spendthrift" attitude);
- The strongest lighting in the zone is taken as the threshold value ("thrifty" attitude: if there is a place with sufficient light, the occupant will move to carry out the task).

The choice of these criteria can be different for supplementary and uniform lighting, but it is defined once and for all, whatever the time and day of the week.

The next step is to define and size the installation. This sizing corresponds to the maximum lighting values requested, and a total absence of daylighting.

b)   Sizing installations

Two types of lighting exist: uniform lighting (for atmosphere), and supplementary lighting (for tasks). These two types of lighting can be obtained by any kind of

---

[333] AFE recommendations, countersigned by tourist amenity managers

[334] It is necessary to determine whether the choice between supplementary and uniform lighting should be exclusive, or whether there is a possibility of giving a value to each at the same time of day

lamp. Atmospheric lighting is uniformly spread out over the surface of the room, thanks to a generally symmetrical layout of lamps. Supplementary lighting does not bring about problems of inter-reflections on the surfaces of the room, and is used to obtain stronger lighting in small areas[335]. These considerations result in three alternatives that make it possible to define and size most artificial lighting installations at building design stage.

c) Lumen method

As for daylighting, calculating all of the inter-reflections on the walls, ceilings and floors is a cumbersome task. The Lumen method for artificial lighting is practical and widely used to model lighting in a simple way[336]. It determines the average lighting E on the work surface with the formula:

$$E = (C_u \times \Phi)/S,$$

where
Φ = total flux emitted by the lamps (without counting absorption by the light fixture) in lm
S = Area of the work surface
C_u = Usability factor.

If all light from the lamps reached the work surface, the lighting would be Φ/S. Thus, the usability factor represents the efficiency with which the room-light fixture combination transfers light towards the work surface. It depends on the room's geometry and the reflection factors of the surfaces and the light. The geometry only intervenes through the index of premises K. The usability factor C_u is supplied in the form of tables by the manufacturers in relation to K and the reflection factors.

Since E is known (lighting required), the necessary luminous flux of the lamps, and so their power P, can be sized thanks to the energy efficiency[337]:

$$\Phi = E \times S/C_u \quad => \quad P = \Phi/e$$

This type of calculation is generally carried out for light fixtures with halogen, compact or fluorescent lamps.

d) Lighting diagram

For light clusters, manufacturers' catalogues provide lighting diagrams giving the width of the light beam and its lighting value in relation to distance. They generally correspond to halogen or compact lamps, and can be used for atmospheric lighting (inbuilt type), or supplementary lighting (spot). In the first case, the beams should meet on the work surface so as to supply continuous minimum lighting on the surface S to be lit.

---

[335] Good design practice recommends that the atmosphere light in a room should not be less than 33% of task lighting, for comfort and easy user adaptation (DOE, 1993)
[336] Rabl Ari and Kreider Jan, *Heating and Cooling of Buildings*, McGraw Hill 1994
[337] It is the designer who then verifies that the room's geometry and the spacing of the light fixtures are suitable to provide sufficiently uniform lighting

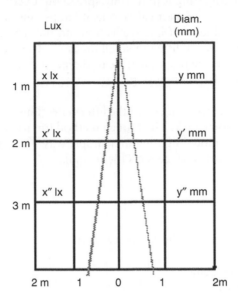

Figure 3.28 Lighting diagram.

The lit surface is given in relation to the distance from the work surface, usually the height H of the lamp minus the height of the work surface, e.g. 85 cm. When choosing a type of lamp, the luminous efficiency gives the number of Watts to supply at a given distance:

$$S_{lamp} = f(H - 0.85)$$
$$n = S/S_{lamp}$$
$$P_{lamp} = E \times S/e$$
$$P = n \times P_{lamp}$$

where S_{lamp} is the surface covered by a lamp, H the height of the ceiling, 0.85 cm the default height of the work surface, n the number of lamps to be placed, and P_{lamp} the power required for a light fixture.

e)   Yield

For many less functional light fixtures, particularly those used in the home, none of these tables or diagrams is provided. Only the notion of light yield has therefore been considered. An overall light yield is then indicated, along with the percentage of the work surface to be lit. As above, the type of lamp chosen makes it possible to calculate the energy consumption:

$$P = (E \times S)/(\eta \times e)$$

where η is the light yield.

Lighting power installed

Table 3.14 gives an idea of the power density installed for artificial lighting. It is taken from the US Department of Energy's building code, which specifies these values (LBL,

Table 3.14 Recommended power levels for artificial lighting (W/m²).

Type of building/ Use of surface area	Size of gross lit surface areas	
	0 to 200 m²	201 to 1000 m²
Office	14	13
Lounge/bar	16	16
School	13	13
Service establishment	28	20

1992). They are given in W/m² in size order, but they actually depend on the type of lamp (and ballast) chosen.

The elements presented in this paragraph thus make it possible to establish a usage scenario with lighting instructions, and then input data on equipment (lamps, light fixtures, settings). The first stage involves calculating the level of artificial lighting on the specified work surface. If this lighting differs from the instructions, the system should be resized. The software then calculates for each time step the daylighting on the work surface, and if this is insufficient, the artificial lighting and thus the electric energy consumed. This energy is considered as an internal gain in the thermal simulation. Seasonal or annual consumption can be deduced by integration over the simulation duration, along with average daylight factors on the work surface.

Notions of comfort

A lighting project needs to satisfy requirements other than simple lighting values. Light plays a considerable psychological role. In particular it has a large influence on how we perceive space (e.g. impression of lightness, space, intimacy), and on our mood. Daylight brings a full spectrum of visible and invisible light. It thus stimulates significant numerous biological mechanisms and gives incomparable colour rendition. In addition, the energy efficiency of daylight is high. Putting aside the problems of undesirable radiation or flashing that may be brought about by artificial lighting, it is preferable to privilege daylighting in a room for reasons of comfort and economy.

Several authors thus recommend minimum daylight factor values, in particular for lighting housing. The table below comes from Bouvier[338]. A daylight factor is equal, for a luminance distribution in a given sky (overcast) at the quotient of the lighting value at point P within the room, to the value of external diffuse horizontal lighting. The daylight factor values obtained are given in percent, with 10% at the centre of a room considered as a high value.

Depending on climate conditions at a given time, the level of lighting obtained, which is only a fraction of the level of lighting in an open site outside, will or will not exceed the level of lighting required for the task. In the latter case, artificial light will be needed. The daylight factor does not therefore determine the level of lighting but the probability of reaching these levels, depending on the local climate. It determines

---

[338]Bouvier François, *Éclairage Naturel*, Techniques de l'Ingénieur, C 3315

Table 3.15 Recommended daylight factors.

Daylight factor at the centre of the room, on a horizontal plane 1 m from the ground		
	Minimum (%)	*Recommended (%)*
Living room	1.5	2 to 3
Bedroom	1	1 to 2
Child's bedroom	1.5	2 to 3
Kitchen	1.5	2

a probable level of energy consumption for lighting, and interior comfort. It is more suited to the early design phases, which does not exclude using a more sophisticated tool to reach a more precise optimum between heat and lighting.

The daylight factor varies depending on the distance from windows. The CSTB has devised a simplified method to estimate the minimum value of this factor in a room lit from one side. Two parameters are defined:

- corrected glazing index, $Ic = t. Sv/Ssol$

where t is the rate of the glazing's light transmission, Sv the surface of the glazing and Ssol the surface of the ground at the premises,

- depth index, $Ip = P/H$

where P is the depth of the premises (in relation to the window) and H the maximum height (under the lintel) of the window in relation to the ground.

The minimum daylight factor Fjm of the room depends on the parameter ($Ic - 5$. Ip) according to the table below.

Table 3.16 Estimation of minimum daylight factor FJM.

Parameter ($Ic - 5. Ip$)	Minimum daylight factor (%)
>5	>2% (very light premises)
from 0 to 5	from 1.5 to 2% (light premises)
from −5 to 0	from 1 to 1.5% (dark premises)
<−5	<1% (very dark premises)

If the premises are lit from two opposite sides, the depth P is half of the distance between these two sides. Additional parameters have been introduced[339]: exposure of the façade (obstruction angle), dirtiness (urban or rural site), distribution of openings, reflection factor of surfaces, and solar protection.

---

[339] Daniel Bernstein, Jean-Pierre Champetier, Loïc hamayon, Ljubica Mudri, Jean-Pierre Traisnel and Thierry Vidal, *Traité de la construction durable*, Ed. Le Moniteur, Paris, 2007

An even simpler (although possibly less precise) calculation[340] is used to make an initial evaluation of daylighting. The daylight factor F_{dj} is given as a % in cases where premises have only vertical glazing:

$$F_{dj}\ (\%) = (\tau \cdot A_w \cdot \theta)/(2.A.(1 - \rho))$$

where τ is the rate of diffuse light transmission through the windows, A_w is the area of the windows, A is the total area of the surfaces in the premises (floor, ceiling, walls including glazing), ρ is the average reflection rate of surfaces in the premises (average of the reflection rates weighted by the surfaces, including glazing), and θ is the angle (in °) of the portion of the sky seen from the middle of the window (cf. figure below).

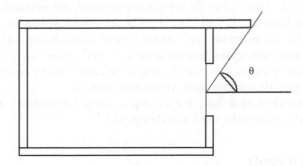

Figure 3.29 Portion of sky seen from a window.

If, for example, the daylight factor is 2%, then the inside lighting will be 100 lux for a typical overcast sky (outside lighting: 5000 lux).

In some specific buildings, especially museums, very strict lighting conditions have been defined so that it is no longer sufficient to quantify the level of lighting; it is also necessary to evaluate the risks of glare, contrasts between different surfaces (e.g. avoid dark surfaces close to windows), colour rendition (which can impose less energy efficient lamps with more appropriate wavelengths). Software exists for this purpose, but it requires very precise input of building data (room by room, indicating the exact position of each window, the properties of each surface, and possibly the furniture).

One software programme available in this area is RADIANCE[341], which uses a ray-tracing method to analyze and visualize light. Taking into account the geometry, materials, natural and artificial lighting, it is possible to calculate the spectra of light intensity (quantity of light passing through a point in a given direction and colour), spectra of total light quantity affecting a surface (quantity of light and colour) and the glare rating. The simulation results can be in the form of images, digital values or curves. A version for personal computers is included in the software programmes ADELINE[342] and ENELIGHT[343].

---

[340] Sumpner W.E., The diffusion of light, Philosophical Magazine, XXXV., 1893 http://www.tandfonline.com/toc/tphm16/35/213#.VR5DiOHK4gU
[341] cf. http://radsite.lbl.gov/radiance/
[342] cf. http://www.iea-adeline.de/
[343] www.izuba.fr and www.deluminaelab.com

3.4 ACOUSTIC CALCULATIONS

In this domain too, simplified tools exist that are suitable for most current situations, along with specialized tools, e.g. to design an auditorium. Acoustic regulations determine minimum performance levels. Depending on the site and the type of building, higher performance levels may be required.

3.4.1 Simplified tools

In most cases, a building's acoustic atmosphere is dealt with by protecting from external noise or noise between different premises (between apartments, between offices and meeting rooms, etc.). Apart from the regulatory method, the evaluation method most frequently used in France is QUALITEL[344]. This method involves evaluating acoustic insulation levels for the different surfaces concerned (considering airborne noise and, for flooring and some staircases, impact noise), as well as the risk of transmission by ventilation and service ducts, woodwork, and noise generated by equipment (e.g. taps, water pipes, heating, air conditioning, ventilation and lifts).

This kind of calculation has, for example, been integrated into the software ACOUBAT, which can thus be used as a design aid[345].

3.4.2 Detailed tools

Other criteria can be considered for some types of space. The reverberation time depends on a space's surface coatings and geometry. It is the time required for a sound to be reduced to a millionth of its initial intensity. A theatre (auditorium for listening to music) must offer sufficient reverberation time: a high-density sound should be audible for one to two seconds after the end of its emission. On the other hand, when reverberation time is too long, it produces marked resonance, even echo. The solution is an appropriate distribution of absorbent and reflecting materials.

Simulation tools divide a building's surfaces into elements: the operator defines the fineness of the mesh depending on the differential acoustic constraints on each surface. Each surface is represented by a polygon, which is given an absorption coefficient. The operator can create a specific mesh per surface, whose fineness will correspond to the analysis precision required for the surface.

Results can include sound levels, the echogram, the decrease in sound in the space, the relative acoustic level on surfaces, the acoustic indications of the room (definition, clarity, lateral fraction), the directivities, i.e. at a given reception point, the directions from which the acoustic density arrives: static directivity (constant sound source) and dynamic directivity (pulsed source), and the temporal distribution of the energy received: instant or cumulated.

Some acoustic engineering tools designed on these bases are available, e.g. MSC ACTRAN, HALL ACOUSTICS, CATT-Acoustic[346].

---

[344] cf. http://www.qualite-logement.org
[345] cf. http://www.cstb.fr
[346] cf. www.planete-acoustique.com, www.mscsoftware.com, www.catt.se

3.5 EVALUATION OF HEALTH IMPACTS

As we saw in paragraph 3.1, life cycle assessments concern overall environmental impacts, but cannot be used to evaluate specific impacts on the health of occupants in the building studied. No evaluation tool currently exists in terms of health to aid professionals design buildings, but publications are available that present the problems and make recommendations[347], as well as databases (in German) on the toxicity and risks associated with different products[348].

Nevertheless, a global method for assessing health impacts is used to evaluate environmental performance (standard ISO 14 031), and this approach could be used in buildings. It is therefore worth presenting it here.

Health aspects in buildings can be grouped into three major domains: aspects linked to air quality, to water quality, and other aspects like the problem of electromagnetic fields. Air quality also influences olfactory comfort, which is why it has been included in this paragraph.

The first stage involves evaluating the quantities of pollutants emitted into air and water, as well as the possible presence of pollutants in the ground. Emissions in the air may come from certain materials (fibres from insulation, volatile organic compounds from paint, glue, chipboard, etc.), some equipment (boilers), activities (cigarette smoke, DIY, maintenance, moving vehicles in a garage, etc.), or the outside atmosphere (radon, urban or traffic pollution, emissions from polluted soil, etc.). Measures are carried out on materials like coatings to ascertain pollutant emissions over time. Emissions in water are linked to the dissolution of metals from pipes (especially lead). The ground on which the building is constructed may have been polluted at an earlier time.

The next step involves evaluating the concentration of the different pollutants in the air and water, depending on their dilution. Water quality may be evaluated according to the characteristics of the drinking water supply (e.g. hardness), the materials used in the building's internal network (metal alloy composition, etc.), the maintenance envisaged (scale removal, anti-corrosion treatment), techniques for preventing legionellosis (temperature of hot water when produced, in the network and in any storage system), and the system for separating drinking and non-drinking water supplies. Depending on the quantity consumed (e.g. 2 litres of water per person per day), the doses received by users can be deduced, and the dose-response functions (cf. below) can then be used to evaluate risk.

The concentration of pollutants in the air depends on ventilation flows from outside to inside the premises, and between different premises. Models exist to quantify these air exchanges in relation to the envelope's proofing (i.e. permeability to air, which can be measured on site once the construction is finished), the ventilation system (air inlets and outlets, a ventilator in the case of controlled mechanical ventilation), and the openings between premises (doors). The circulating airflows depend on local weather conditions (wind direction and speed), and the outside and inside temperatures, which condition the chimney effect (hot air rises because it is lighter). These conditions are

[347] Suzanne and Pierre Déoux, *Le guide de l'habitat sain*, Ed. Medieco, 2002
[348] CD ROM ECOBIS, *Okologisches Baustoffinformationssystem*, German ministry for transport, construction and housing, June 2000

often difficult to evaluate because of very localized disturbances, for instance caused by neighbouring buildings, in which case these local variables cannot be precisely ascertained. Once the emissions and ventilation flows are known, even approximately, the air concentrations can be deduced.

Health effects are not directly linked to the concentration of pollutants in the air, but to the doses received by the people exposed to them. Doses are expressed in kg (or mg) of pollutant inhaled or ingested per kg of human weight. These doses depend on concentration, as well as the quantity of air (or water) ingested (on average an male adult inhales $23\,m^3$ of air per day, a woman $21\,m^3$ and a child $15\,m^3$), the person's weight (at equal concentration, the dose is generally higher for children), and the exposure time (the dose is higher if the pollutant is emitted in a bedroom, where people spend more time than in a corridor, for example).

Health effects can then be estimated using the "dose-response" method. This method is based on databases that link the number of diseases generated to the doses received, for a certain number of substances. For example, chloroform induces a 0.081 risk of cancer (8.1% probability) for a chronic daily dosage (i.e. over a person's whole life) of 1 mg/kg[349]. This kind of data is unfortunately only available for a fairly low number of substances (around one hundred) out of the thousands of chemical substances on the market. In addition, interactions between substances are not taken into account. For carcinogenicity, the response is proportional to the dose taking it that no threshold exists: even a very small quantity of pollutant can have consequences. For some other illnesses, however, a threshold may exist: if the dose is below a certain limit, the pollutant can be assimilated by the organism without any detectable health effect.

Global indicators are proposed based on the probabilities of generating a certain number of diseases. An initial approach involves putting a price tag on health impacts: the average cost of care corresponding to each disease must be estimated, along with the cost of loss of human life (which raises a number of philosophical and ethical questions). Another approach consists in evaluating the number of years of human life lost and the number of years of life in good health lost (subject to knowing the average duration of the diseases considered), cf. the DALY indicator presented in chapter 2.

This presentation illustrates the complexity of establishing a comprehensive evaluation of health risks, in particular in buildings. This is probably why this issue is currently dealt with by an expertise approach rather than by calculations like the domains covered above.

3.6 WORKSITE WASTE MANAGEMENT

A tool has been developed in France to aid worksite waste management: ECO-LIVE[350]. Users need to input data on the worksite (construction or demolition, location, space available for skips) and the building (type, surface, detailed description by parcel, materials used for foundations, surroundings, facades, roof, flooring, ceilings and inside walls, windows, stairs, surfaces, etc.).

---

[349]US Environmental Protection Agency (EPA), Health effect assessment summary table (HEAST), report EPA/540/1-92/002, Washington, January 1992
[350]cf. http://www.recy.net/outils/adatire/adatire.php

The software evaluates the quantity of waste grouped into categories (inert waste, banal or dangerous waste). The cost of processing is then estimated depending on the choices of treatment, which can be compared.

3.7 GENERAL TOOLS

No tool currently exists covering all of the evaluations presented above. Some general tools have been developed.

The longest-standing method is BREEAM, devised by the Building Research Establishment in the United Kingdom[351]. It works on a principle of granting credits to projects based on performance criteria (energy requirements, CO_2 emissions) or techniques (e.g. presence of an outbuilding for bicycles, an extractor fan over a stove, etc.). The number of credits allotted at each performance level and for each technique has not been justified by an environmental balance sheet, which would no doubt be difficult on a general level. The method's main strong point is that it constitutes a checklist of the key factors in designing or evaluating projects. Its simplicity also makes it popular.

Methods like PAPOOSE, developed by B.E.T. TRIBU, or ESCALE, developed by the CSTB[352], comprise a number of evaluation modules for different criteria, but these evaluations are relatively simple. If the sensitivity of a simplified model is insufficient, then the design choices can result in mediocre performance. Thus, in a "high environmental quality" college located in Ile de France, the energy consumption for heating is over 50% higher than average for a new college, despite the fact that the project had been positively evaluated by a general method. Developers are constantly improving methods of this type and we can expect greater precision in the future. They are currently not generally available, but are used by their creators for projects, in particular to provide assistance to project managers in writing programmes.

The SB Tool (formerly GB Tool) developed by an international group as part of the Green Building Challenge[353] is another general tool that integrates environmental indicators (e.g. potential for global warming and acidification), qualitative criteria (green space, building's adaptability), and technical criteria (presence of air conditioning, heat recovery on ventilation). But the evaluation methods of each criterion are not described: e.g. the global warming potential of the project studied must be compared to a reference value, and a grade ranging from -2 to $+5$ is deduced. A life cycle assessment could be used for this evaluation, but it is not included in the software and thus users often settle for an "expert opinion" evaluation, which is subjective to say the least.

Taking another example, summer comfort, a low grade is obtained if the humidity rate is not controlled, such as when there is no air conditioning. However, in most French climates the humidity rate rarely poses a problem, and occupants may judge a building that is passively cooled to be very comfortable, despite the fact that it would not correspond to an evaluation using the SB Tool. Other contradictions of this type, linked to specific regional features (can a tool be valid for all countries and

---

[351] www.breeam.org
[352] Rialhe A. and Nibel S., *Quatre outils français d'analyse de la qualité environnementale des bâtiments*, Ed. Plan Urbanisme Construction et Architecture, 1999
[353] International Initiative for a Sustainable Built Environment, www.iisbe.org

all climates?) simplified methods ("expert opinions", addition of credits), and value judgements implicit in evaluations (criteria weighting, criteria structure), mean that great care should be taken when using this tool, which responds to an objective of raising international awareness but does not constitute a genuine design aid.

AFNOR is currently drawing up a standard to evaluate buildings' environmental quality[354]. The experimental nature of this text means that tools can progress in line with available knowledge.

The general tools currently available respond to short-term requirements, often resulting from more subjective than technical approaches. They do not always result in reducing buildings' environmental impacts. Users, contractors and project managers should therefore take a critical approach when interpreting results obtained using these tools.

3.8 USING TOOLS

3.8.1 Programming

3.8.1.1 Retrofitting or (re)construction

The choice of whether to (re)construct or retrofit is one of the first alternatives facing contractors. Retrofitting may initially appear more advantageous. However, it is important to assess impacts on the building's life cycle and not just to compare the respective impacts of retrofit and construction worksites. This involves in particular taking the energy consumption of both "systems" into account, along with replacement of components over the whole life span. A comparative life cycle assessment can be carried out.

The results depend on the final quality of the renovated building, particularly from a thermal point of view, but also the reduction of water consumption and the waste sorting potential (e.g. enough space in the building's kitchens and service rooms).

3.8.1.2 Choice of site

When comparing sites numerous parameters should be considered:

- Energy aspects: orientation in relation to the south and presence of masks (other buildings, trees, etc.), exposure to wind, connection to gas supply and district heating networks (heating networks that recover geothermal or wood fuel energy are environmentally interesting), additional public lighting required by the project;
- Land use: unspoilt site or reconstruction, presence of historic monument or conservation area, impact on vegetation (cutting down trees), previous ground pollution;
- Air quality: presence of industrial or road equipment upstream from dominant winds, presence of radon;

---

[354]AFNOR, standard NF P01-020-1, Bâtiment – Qualité environnementale des bâtiments – Partie 1: cadre méthodologique pour la description et la caractérisation des performances environnementales et sanitaires des bâtiments, March 2005 (the other parts are being drawn up: a usage guide in part 2, and an experimental standard defining performance indicators in part 3)

- Noise: presence of a close source (road, railway, industry, etc.);
- Waste management: treatment mode (landfill, incineration with or without energy recovery/cogeneration, compost), selective collection for different materials (glass, paper, packaging, etc.) and distances from treatment sites;
- Transport: public transport services, travel distances (work-home, shops, etc.), bicycle path;
- Water management: composition of water supply (nitrates, sulphates, heavy metals, pesticides, hardness), supply yield, sewage connection rate;
- Risks: landslide, seismic risk, flooding in the zone (flood risk and water table exposure), proximity to forests and fire risks in some regions.

These different aspects condition the impacts over the future building's lifetime, and its internal environmental quality in terms of comfort and health. A life cycle assessment can be carried out on a typical building (individual house, community housing, tertiary building, school, etc.) prior to any architectural design, and for a typical occupancy scenario (quantity of waste produced per day, quantity of water consumed, desirable temperature for heating and possibly air conditioning, energy consumed for hot water, cooking and lighting, daily transport used). This analysis quantifies the impacts for each site, using an environmental profile comprising several criteria. Depending on local priorities (e.g. the problem of water resources may be locally more important than e.g. acid rain), a preference may be given to one of the sites.

However, some interactions between the site and the building need to be considered. For example, proximity to the water table may result in a decision to increase the mass of the foundations (tie beams may also be used against buoyant force) or avoid building basements (with other potential impacts like land usage for car parks).

3.8.1.3 *Drawing up the architectural programme*

During this stage, designers are informed of the client's requirements, in particular regarding the functionalities of the building. This is essential because environmental quality results from a good match between the architectural design and equipment, and the anticipated uses. These uses should therefore be properly assessed. For example, the quantity of hot water required in a community facility (e.g. school, hospital) conditions the size of the production system. If the requirements have been wrongly evaluated, the sizing will be incorrect, which can lead to a drop in the yield (especially for solar installations).

Once the requirements have been carefully defined, the programme can direct the designers and producers towards objectives or specific measures. For instance, it may recommend a life cycle cost calculation for a system. Targets can be specified[355], e.g. reducing water consumption using appropriate equipment, or energy demand with a bioclimatic approach and efficient equipment, managing waste in line with municipal options (sorting space in kitchens and technical rooms).

---

[355] Call for offers by the French construction and architecture programme on high environmental quality experimental buildings, November 1993

Recommendations may be included in the programme regarding the orientation of rooms depending on their use, the position of bay windows, protection against overheating, thermal mass, ventilation, etc. Environmental quality should not however impinge on architectural creativity. This approach applies to all architectural styles and does not justify imposing fixed technical solutions.

Taking the example of heat, the programme can specify performance based on heating load (e.g. below $15 \, kWh/m^2/year$) or summer temperatures[356] (e.g. maintain max. 27°C in main rooms when average daily temperature does not exceed 24°C). Environmental performance is sometimes stipulated, as in the renovation of the Lyon Confluence neighbourhood (cf. chapter 5), for which thresholds have been fixed for CO_2 emissions and radioactive waste generation.

3.8.2   Project management, architectural design and engineering

The order in which designers consider the various quality aspects will affect the overall performance of a project: taking care of aesthetic and functional constraints and then verifying whether they match minimum regulations on environmental quality will usually give a different result from an approach that integrates environmental considerations right from the start.

One of the early decisions concerns how the project fits in with its environment (networks, neighbourhood, etc.). Initial attempts to produce ecological buildings have sometimes involved trying to make them self-sufficient, for water (recuperating rain water, local collection), energy (inter-seasonal storage) and waste (compost and local processing). In reality, this may not always be the best choice. Creating zero emissions over a system's lifespan (one of the objectives of self-sufficiency) imposes making an "environmental investment" in construction, which is not necessarily profitable in the long term. Taking heat as an example, building a seasonal storage system for a detached house requires a significant amount of materials. Even less expensive hydrogen storage does not have a positive life cycle balance compared to a house with low requirements (bioclimatic or passive design)[357]. A house that is well integrated into networks can take advantage of efficient equipment (e.g. water purification, energy production and supply, waste management), which can often be advantageous from an environmental point of view.

Sharing roles between individuals and the community is not however easy on a wide scale. Network leakage, the effects of storms and frost, greater flexibility of controls for individuals, and making users responsible through an individual cost system all argue in favour of self-sufficiency, or at least individual systems. The intermediary scale of neighbourhoods or urban islands can make it easier to use alternative techniques and take advantage of local resources (e.g. small heat network fuelled by a wood boiler, a mini electricity network fuelled by solar and/or wind equipment, micro-cogeneration, etc.).

In terms of energy, making use of solar power, clean and renewable energy at local level is often cheaper than using centralised facilities like solar power stations:

---

[356] M. Gerber and ADEME Languedoc-Roussillon, *Cahier des charges à l'usage des concepteurs*, October 1991

[357] German Engineers Association (VDI), Munich conference, December 1993

solar collectors replace existing components (e.g. part of the roof), which avoids both expense and additional land use. A building offers considerable collection systems (façades and roof). Over the heating season, a well-positioned house located in Ile de France at current standards (RT 2005) receives (on its roof and façades) an amount of solar energy equal to twelve times its heating load (annual solar radiation on a horizontal plane is over $1,000 \, \text{kWh/m}^2$). Yet in addition to collecting the energy, it needs to be transformed into heating savings. This requires three successive transformations.

- Capture, transformation of light into heat: solar radiation is absorbed by a surface which heats up and can yield heat;
- Storage, phase difference in solar gains: heat must be partially stored from day over night using thermal mass (heavy masonry), and then distributed at the right moment; however, inter-seasonal storage is only profitable and efficient for large systems that feed into several hundred buildings (small storage systems comprise more loss surface in relation to their volume and conserve heat less over time);
- Distribution and control: heat must be distributed in the premises according to needs, and control must follow solar gains (e.g. using thermostatic taps for water installations), otherwise the result will be overheating instead of lower consumption; inert space-heating systems (e.g. under-floor heating) are less suitable than, e.g. air systems; in addition, the generator's yield must remain as high as possible at partial load, since solar gains reduce the power demand in relation to the nominal load (maximum power of the boiler, calculated for the coldest days with no sun).

However, significant recovery of solar gains requires anticipating how the building will behave in summer, and also in mid-season, and avoiding overheating using suitable means: solar protection (e.g. roof overhangs that block rays in summer when the sun is high, blinds, plants, etc.), thermal mass (possibly use of the ground), ventilation (in particular for sunspaces).

The only solution for effective energy savings without reduced heat comfort is early collaboration between architects and engineers. The COMFIE software's energy analysis using multi-zone simulation has been specially designed to fulfil these objectives. The dynamic analysis makes it possible to obtain temperature profiles in the different zones of a building and evaluate annual heating loads. It is then possible to compare design alternatives to limit requirements while keeping down the risk of overheating.

A life cycle assessment can be used to generalize the energy approach by considering the numerous environmental effects described in chapter 1. Data on a building's geometry and components along with the energy requirements calculated by the COMFIE software serve as an entry for the environmental quality evaluation tool EQUER. Some additional characteristics are required, in particular:

- type of site (urban, suburbs, rural, remote);
- lifespan of the building, windows, surface coatings (suggested default value respectively: 80 years, 30 years, 10 years);
- surplus factor for the worksite (default value: 10%);
- distances for transporting materials (default value: 100 km);
- type of energy for heating and hot water (gas, fuel, electricity, wood, district heating);

- consumption of hot and cold water (default value: 40 and 140 litres per person per day);
- household waste generation, sorting, collection and treatment (default value: 1 kg per person per day);
- transport distances (home to place of work, shops, public transport system), default values depending on type of site.

The number of these parameters has been limited thanks to the use of default values (e.g. the temperature of hot water is 55°C, the cold water temperature is estimated according to the latitude). These default values usually have a low impact on the overall result, and the user can modify them manually in the data files (in the tool's research version).

Other technical evaluation tools concern acoustics and lighting. They can be used to evaluate the level of acoustic and visual comfort, along with provisional consumption for artificial lighting, taking into account the contribution of daylighting.

These technical analyses should not of course hamper the architecture's aesthetic aspects. A few theoretical attempts have been made, but a counter-example can be found for most affirmations on the subject. Aesthetics are often associated to a uniquely visual image, but studies have been carried out on links with acoustic and even thermal aspects (e.g. coolness and acoustic reverberation can reinforce the aesthetics of some locations). The same architectural project can be done in various ways, each expressing a different atmosphere (intimacy, ceremony, fantasy, etc.). The image of a building or premises can have significant consequences on an activity (e.g. expressing the wealth of a banking or tertiary establishment, etc.).

The image and/or atmosphere sought can be obtained by taking a deliberate design approach following a series of stages, for instance[358]: the programme (defining the context and objectives by asking the project manager questions to determine whether the project between open and closed, more institutional or family-based, luxury or simple, etc. and establish the list of qualifiers), collecting examples of existing buildings, defining alternatives corresponding to the objectives sought, and validating alternatives in relation to other constraints. Formalizing the aesthetic approach in this way facilitates communication in the design-production team, thus making it easier to fulfil objectives. The link between subjective effects and technical resources (materials, decoration, style of letters and icons, partitioning, plants, lighting, etc.) remains a vast field of research that is still little explored. It is important to emphasize that integrating environmental quality objectives does not necessarily modify the image and atmosphere of a building, unless the project manager wants to put forward a "green" image or is seeking to optimize performance (in the actual state of the art) without any particular aesthetic requirement.

3.8.3 Managing a worksite

Although the relative impact of a building worksite is low in relation to the impact of a building over its lifespan, it is worth reducing nuisances (noise, dust, waste generation)

---

[358]Peter Manning, *Environmental aesthetic design*, Building and Environment vol. 26 No. 4, 1991

that can create discomfort for neighbours. These nuisances are more visible than the consequences of prevention actions, for example careful insulation (in particular reducing thermal bridges), and controlling air infiltrations (proofing), which are often very beneficial to environmental quality. It is important to concentrate not just on the spectacular aspects of worksites' environmental quality, but to think in terms of the whole life cycle.

Good management helps reduce waste matter, with the advantage of combining environmental quality and economic profitability. Unused materials (e.g. impaired items, remains, off-cuts, etc.) can represent 10% of total weight, but this figure can vary significantly from one worksite to another (more intensive use of surface calculations can bring down surpluses) and depending on materials. Waste management regulations should be followed when getting rid of surplus concrete. Packaging also constitutes a significant proportion of worksite waste. Unregulated incineration, which is still frequent practice, generates emissions with impacts on health, the greenhouse effect and acidification. Transportation to a waste treatment installation should be included in the services proposed to the ordering party. Containers can be provided for sorting worksite waste: wood, metal, inert waste (dry cement and other rubble), plastic packaging and cardboard. Surplus paint and other dangerous waste matter should also be collected.

Particular care should be taken in treating used oil, solvents and other organic products, particularly on worksites located close to the water table. Vegetable oils can be used to reduce the impact of concrete formwork. Oil from worksite machines should not be drained on site, but in suitable locations. A plan to manage rainwater around the site should be put in place to limit the risk of pollution (drainage and removal). A zone covered with a tarpaulin can be set up to carry out mixing (mortar, concrete, etc.).

Various measures can reduce noise on the worksite. For example, self-consolidating concrete avoids the use of vibrators. In urban zones, sometimes trees must be preserved on the terrain.

One option is to re-use[359] components (e.g. woodwork, material and plumbing), but this is rare. Recycled materials are increasingly used (e.g. pipes made from recycled PVC, replacing a proportion, e.g. 20%, of aggregates with concrete chips for raft foundations). The steel employed in buildings already contains 50% of recycled material. Another way of reducing the impacts of construction is to reduce consumption (e.g. use thinner concrete reinforcements).

Regarding air quality, it is preferable to treat wood with products that contain no lindane, pentachlorophenol or other organochlorines (xylo…). The reason is that these products are harmful and progressively release treated wood over a long period. Pyrethrinoids (cypermethrin, permethrin), triazole, boron salts and distillates of charcoal appear to be less harmful[360]. It is also important to make sure that occupied spaces are properly separated from fibrous insulation (suspended ceilings, proofed wall linings, etc.) to reduce the amount of fibres in the air. Toxic emissions can be minimized by

[359] Re-use is when a component is used again for the same function, whereas reclamation involves a change of function (e.g. used tiles can be used as a garden border or to cover a different roof)
[360] Que Choisir, *Guide pratique des produits propres*, Nov./Dec. 1990

using paint or solvents that release few VOCs, solvent-free adhesives, and chipboard containing lower levels of formaldehyde.

The overall quality can be improved by coordinating the various professions. In particular, air infiltrations concern the structural work, the woodwork, and some of the electric equipment built into the walls. Different professionals thus need to intervene in a consistent way. The design team must give details, such as:

- to plumbers, on energy-saving sanitary equipment;
- to heating specialists, on fitting installations with suitable power (too large means a reduced load and a lower yield) along with suitable control of the heat and ventilation system. Making energy savings by reducing the air renewal rate is not recommended, since it affects air quality. Controlling airflow in line with occupancy is a better solution, e.g. using humidity-sensitive ventilation;
- to electricians, on making sure plugs are properly earthed, and keeping wiring away from beds;
- to painters, on choosing particular products (NF-Environment standard).

The choice of building materials appears to play a relatively insignificant role in the overall environmental impact of a standard building's lifespan. However, this is no longer the case for high-performance buildings, cf. the balance sheet of a passive house presented in chapter 5. This choice should include not only the environmental quality when manufacturing products, but also their durability and the emissions they generate over their lifespan. Paint that generates 30% more VOCs that another paint when manufactured but that needs replacing half as often usually presents a positive balance sheet over its lifespan.

Uncertainty and lack of current data on manufacture life cycle inventories for products and construction work make it difficult to take a rational approach and, except for extreme cases (e.g. asbestos), it appears preferable to base decisions on products' functionalities rather than partial environmental quality (e.g. only based on analyzing impacts in the manufacturing phase). When choosing between two similar products, the priority should be on the one with an environmental advantage (e.g. higher proportion of recycled material, production process with lower impact, etc.). Some technologies are presented in chapter 4.

The acceptance of completed work is an essential phase that is often neglected. The designers should be present when the work is accepted to verify that everything has been correctly carried out.

3.8.4 Managing housing stock

User behaviour has as much influence as design choices on final performance. Behavioural aspects generally explain the gaps observed between anticipated performance and the performance measured on site, e.g. the heating consumption measured (and invoiced) depends on thermostat settings, which can be different from calculation hypotheses (usually 20°C in housing).

In order to better comprehend the relationship between design and usage, a double sensitivity study was carried out based on the life cycle assessment. The comparison focuses on:

- two types of building, a reference corresponding to a standard individual house (RT 2000) in Ile de France, but facing north, and a bioclimatic house (rated "HEQ" in the table below),
- two extreme behaviour types: a "thrifty" occupant and a "spendthrift" occupant.

The different parameters are summarized in the table below:

Table 3.17 Data on sandard and "HEQ" construction.

Component	Reference (Ref)	"HEQ"
Insulation	8 cm internal	12 cm external
Glazed surface	10 m²; north-facing	25 m²; south-facing
CMV	no exchanger	dual flow; efficiency 0.5
sanitary equipment	standard	flow reduced by 50%
waste sorting	glass only	glass and paper

The occupants' behaviour is characterized by a number of parameters regarding water, energy and waste management. The table below gives a comparison of the two scenarios.

Table 3.18 Data concerning "thrifty" and "spendthrift" occupancy.

Parameters	"Thrifty"	"Spendthrift"
Temperature setting	Varies from 14°C to 19°C	Constant 21°C
Ventilation	0.5 volume per hour	1 volume per hour
Specific electricity	150 W	300 W
Hot water	40 l/pers/day (1)	60 l/pers/day (1)
Cold water	80 l/pers/day (1)	150 l/pers/day (1)
Household waste	0.8 kg/person/day	1.5 kg/person/day
Paper recycling	60% (2)	0%
Glass recycling	80%	0%

(1) this flow is halved in the "HEQ" alternative
(2) 0% in the "reference" alternative

The results are expressed in the form of indicators evaluated by the EQUER software programme. These indicators are represented as a relative value in relation to the alternative generating the most impact (cf. figure below, indicators vary from 0 – no impact – to 1 – maximum impact). The results of this dual comparison help us answer an important question: is an occupant's "spendthrift" behaviour likely to compromise designers' efforts to obtain high environmental quality (HEQ) housing?

Thankfully, the answer is no! It does confirm an observation made elsewhere, i.e. that user behaviour has a strong influence on the consumption of housing and by extension on its environmental impact; however, this does not override designers' efforts to propose environmentally efficient buildings.

Overall, a "spendthrift" occupant of HEQ housing creates very little more impact than a "thrifty" occupant of reference housing, apart from household waste

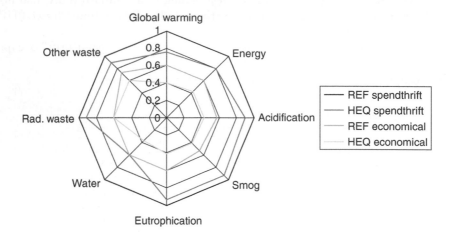

Figure 3.30 Crossed influence of design and occupant behaviour.

management. However, a "thrifty" occupant will take greater advantage of an HEQ house and will end up with a particularly low impact profile.

This comparison therefore fully confirms that HEQ buildings must be designed even when environmental awareness of future occupants is not known (as is usually the case). It also shows that it is preferable to give users very detailed information on the various equipment available to them, and in particular its optimum usage: heating and ventilation facilities, water supply and devices, waste collection system, etc.

Instructions on how to use the building can be very useful for encouraging occupants to manage environmental quality in the long term. Heating requirements are highly dependent on the chosen temperature setting. A temperature of 19°C is usually sufficient, and can even be lowered to 17°C in bedrooms. In cases of irregular occupancy, hour-based settings result in higher savings, and instructions for thermostats and thermostatic taps are always worth providing.

The "robustness" of a system determines the maintenance of its performance independently from the occupants' behaviour. To be really efficient, high environmental performance systems should generate the least possible constraints for users. For example, a sunspace built on to a house can be designed to limit the impact of opening the door that leads from the heated area (the housing) to the buffer area (the sunspace). The most "robust" design consists in pre-heating in the sunspace the fresh air arriving in the house. Air inlet grids are therefore fitted between the sunspace and the outside, and at the top of the door between the sunspace and the house to allow the air to pass through even when it is closed. If the occupant leaves the door open, the depression created inside the house by the air extraction (controlled mechanic ventilation) tends to reduce "reverse" air circulation from the house to the sunspace, thus limiting the heat loss. The house's instruction manual can also show the savings to be made by keeping the door shut in wintertime.

The choice of domestic appliances determines some types of consumption: electricity consumption for lighting a room can vary by a factor of 10 from a halogen lamp to an energy-saving lamp; energy-saving (but not water-saving) refrigerators, dishwashers and laundry washing machines are also available. It is now possible to buy "green" electricity kWh: for a slightly higher price, they are guaranteed to have been produced using less polluting techniques, like renewable energy[361].

Saving water is also largely dependent on user behaviour, both for kitchens and bathrooms. One option is to make occupants more aware about the quantity of water they use and leaks (a leaking tap can consume up to 300 litres/day, a flush system 500 litres/day), as well as the quantity of products (e.g. detergent) required in line with water hardness (cf. chapter 4). The building can also be cleaned with less harmful products.

Another very important aspect of eco-management is sorting household waste. To make this easier, suitable devices are studied in kitchen equipment. However, they will be of no use if containers are not regularly emptied and transmitted to collection operators. This activity should therefore be anticipated in the maintenance and operation of the building.

Lastly, it is much easier to repair and retrofit a building if information is kept on the materials and components. This also facilitates diagnosis in case of anomaly. For instance, abnormally high heating consumption can be explained by a dysfunctional boiler or worn thermal insulation. Information on the building's characteristics provides keys that can lead to finding an adequate retrofit solution. Maintaining and replacing components and equipment helps improve environmental quality, in particular of boilers. A third of French individual water heaters (in main residences) are over 15 years old. These old water heaters are a lot less efficient than new apparatus, and their yield drops with age (limescale, etc.).

3.9 CONCLUSIONS AND PERSPECTIVES FOR TOOLS

Among the recurrent problems that dissuade professionals from using tools are: the difficulty of linking different thermal, lighting, acoustic and LCA models, applying them to the first stages of the project (rough plan), and applying them to sizing rather than just performance evaluation.

3.9.1 Inter-model linking

Data input is a particularly high constraint in the building sector, where even small-scale projects include a high number of surfaces, windows and volumes (rooms, etc.). It would be practical to input a project in one go and use a single model for all evaluations (heat, lighting, acoustics, life cycle, cost assessment, etc.). New computer-assisted

---

[361] For example www.enercoop.fr

design tools are object-oriented[362], which makes it easier to link these design tools with technical calculations. Several difficulties exist however:

- Each technical domain requires a different building representation (e.g. rooms for lighting, housing for acoustics, "zones" for heat);
- Objects usually have different names in databases (e.g. a composite surface has a given name in a cost base, it has to be broken down into different materials for heat calculations, with different names in a life cycle assessment base);
- Technical calculations require more data than drawings, and this information must be included in the object model common to the different models, otherwise it will be lost when data is exchanged.

The advances made in object-oriented modelling and model standardization bring the hope that general tools will be developed in the future.

3.9.2 Adaptating models to project progress

The decisions that have the biggest consequences on a building's environmental quality are generally made during the first phases of a project. For instance, the plan of a ground parcel and the design sketch play a major role in future energy consumption. However, many data required for technical calculations are not yet defined at these points. This argument has frequently been used to propose so-called "simplified" methods, which often result in false solutions because they do not make it possible to carry out the sensitivity analyses needed for the project's progress, and merely back up predictable results that do not require calculations, or even assumptions.

It is in fact possible to use detailed models considering "default" values. For example, if the composition of a surface is unknown at the sketch stage, a standard composition representative of current or envisaged construction may be proposed by default. A sensitivity analysis can then be used to compare alternatives (e.g. internal or external insulation) to direct the progress of the detailed project.

3.9.3 From evaluation to design

In the design process, the objective is not so much to evaluate a project's performance as to size certain objects and choose between different technologies. Yet knowledge-based models are oriented more towards evaluation. For example, to size an overhang above a bay window, the current method consists in varying its geometry and choosing dimensions according to the results obtained on heating loads, possibly air conditioning, and comfort levels. This kind of sensitivity study makes it possible to choose a satisfying compromise, and the architectural mask can be supplemented by a shading device if necessary.

It would be useful to have a tool to resolve the opposite problem: i.e. start with a given performance in terms of heating and comfort requirements, and obtain the overhang's corresponding geometry. The problem is more complex than it appears, since

---

[362]Cf. Studies by the International Alliance for Interoperability, which result in standards for data exchange between software, called "Industry Foundation Classes"

the compromise depends on the relative weight given to the different criteria (comfort, architectural quality, and energy saving in the above case, but the environmental quality criteria can be more numerous). In addition, the problem is non-linear and analytically unresolved, which makes an opposite resolution very difficult. Lastly, on a practical level, the solution is not unique, which fortunately avoids having to apply an undesirable degree of architectural uniformity. Tools should thus remain design aids for improving a project while respecting the architects' style.

Despite these limitations, existing tools merit more use by practitioners, some of whom all too often base their decisions on habit or assumptions. Lists of good or bad materials, ratings methods for different techniques, and general but superficial tools may be attractive initially, but their limitations rapidly emerge when used. Current user interfaces make it easier to use more precise software within the time limits imposed by the financial constraints of studies. The objective of these tools is to enrich the design process by facilitating exchange of information on the proposals of the various actors and their evaluation. They make it possible to compare design alternatives, and so genuinely work as a team on a project.

Clients can use tools as well as designers. Tools can help improve the quality of constructions, but they will only be used by designers if the programmes require precise levels of quality. The indicators presented in chapter 2 and the tools presented in this chapter show that it is possible to stipulate a number of thresholds in a programme, e.g.: energy consumption limits (heating, lighting, in $kWh/m^2/year$), daylight factor thresholds for daylighting, temperature thresholds for summer comfort. These limits can be checked afterwards by measurements. Some programmes may also require the use of methodological tools like LCAs to support some technical choices, e.g.:

- Reducing quantities of materials when thermal mass is not necessary;
- Optimizing insulation thickness or glazed areas according to environmental parameters at equal comfort levels
- Choosing between solutions resulting in different heating and lighting requirements (e.g. windows facing more northwards or southwards), based on more general environmental criteria such as preserving resources or protecting the climate.

One potential approach is to bring together all stakeholders in a project (client, architect, engineering firm, contractors, future occupants, neighbours, etc.) to establish a list of quality criteria and priorities. This list can then be part of the programme and act as a guideline for later stages in the project (design, production, delivery, management). This project management principle is described more fully in chapter 5.

3.9.4 Designing neighbourhoods

The tools described above apply to buildings, but the choices made at neighbourhood level (e.g. layout of roads, master plan, community facilities) have a significant influence on a building's performance. The European Eco-housing[363] project, and

---

[363] ECO-HOUSING, Environmental co-housing in Europe, European project number: NNE5/2001/551, publishable final report, January 2006, 29p

the ADEQUA[364] project in France have therefore extended the life cycle assessment to neighbourhood level. The method is presented in chapter 5 in the form of an application performed in the Lyon Confluence neighbourhood.

Other tools have been developed focusing on urban level for acoustics, air quality and external atmosphere quality[365]. A multi-criteria method has been put together based on these different evaluations[366]. Studies continue in this domain and could be useful for the numerous eco-neighbourhood projects currently underway.

---

[364]http://rp.urbanisme.equipement.gouv.fr/puca/activites/rapport-amenagement-durable-des-quartiers.pdf
[365]For example the software SOUNDPLAN, www.soundplan.com, and studies done by CERMA at the Nantes school of architecture, www.cerma.archi.fr
[366]Frédéric Cherqui, *Méthodologie d'évaluation d'un projet d'aménagement durable d'un quartier – méthode ADEQUA*, Doctoral thesis, La Rochelle University, December 2005

Chapter 4

Building "eco-techniques"

The building and public works sector directly contributes to the environmental impact of human activity because of the significant quantity of resources it uses (building materials, energy, water) and the corresponding pollution (waste, smoke, waste water). It also makes an indirect contribution by possibly adding transport requirements (e.g. building a tertiary building far from public transport facilities), curbing waste sorting (e.g. a residential building without sufficient space for selective collection), and increasing electricity demand (extension of urban lighting, electric heating or heat pumps), etc. The "eco-techniques" considered in this chapter are therefore not restricted to construction techniques. Managing flows of matter and energy is also crucial.

To give an idea of size, table 4.1[367] shows pollution emissions in the air in 1991 for the residential and tertiary sector (thousands of tonnes). In addition, a significant proportion of emissions come from thermal power plants (used at peak times for electricity heating)[368] and gas distribution, as well as some of emissions from extracting fossil fuels and at refineries. Waste treatment also plays a role, and some industrial impacts should be taken into account (for manufacturing building materials and components: e.g. cement works are responsible for 14% of industrial CO_2 emissions[369]), as well as a share of road and rail transport due to the choice of residential and tertiary sites.

The building sector represents over 40% of energy consumption in France (around 70 million TEO per year in final energy in 2005[370]) and 23% of greenhouse gas emissions. One m^2 of housing is responsible for emitting around 2 tonnes of CO_2 over its total lifespan (all phases, from manufacturing materials to demolition)[371]. The thermal regulations of 1974, 1982, 1988, 2000 and 2005 aimed to reduce provisional heating consumption by respectively 25%, 20%, 25%, 20% (in the residential sector, 40% in tertiary, which was not regulated before) and 15%[372], i.e. a cumulated reduction of 70% in residential and 50% in tertiary.

---

[367] Serge Lambert et al., *Manuel environnement à l'usage des industriels*, AFNOR, 1994

[368] Source CITEPA, Documentary study No. 99, January 1991

[369] Ministry for the economy, industry and employment, Department of industrial studies and statistics, Building materials, www.industrie.gouv.fr/sessi/publications/etudes/mat/materiaux.htm, 2008

[370] *Les chiffres clés du bâtiment*, ADEME, Ed. 2006, www.ademe.fr

[371] Eric Labouze, *Bâtir avec l'environnement, enjeux écologiques et initiatives industrielles*, Ed. de l'Entrepreneur, Paris, 1993

[372] www.rt-batiment.fr and www.logement.gouv.fr

Table 4.1 Principal flows of material per sector (France, 1991).

| | Local level | | | Regional level | | | Global level | | | |
| | | | | Acid rain | | Precursors photochemical pollution | Indirect greenhouse effect | | Direct greenhouse effect | |
Sector	PS[1] kt	Pb kt	NH₃ kt	HCl kt	SO₂ kt	NOₓ kt	NM HC[2] kt	CO kt	CH₄ kt	CO₂ Mt
residential/ tertiary	15	–	–	10	190	80	140	1000	55	25
– district heat.	3	–	–	2	49	12	1	100	–	1
– home heating	11	–	–	6	137	65	41	900	54	23
– home activit.	–	–	–	–	–	–	97	–	–	–
energy										
– foss. extract.	–	–	–	–	–	–	–	–	214	–
– refineries	9	–	–	–	113	17	66	100	–	4
– gas distrib.	–	–	–	–	–	–	–	–	131	–
– elec. prod.	35	–	–	18	374	121	–	300	–	12
waste treat.										
– landfills	–	–	–	–	–	–	1	–	3436	–
– incinerators	–	–	–	33	–	19	–	10	–	–
industry	150	1	20	15	460	200	525	1300	5	30
transport										
– road (petrol)	20	4	5	3	93	627	1626	9500	79	16
– road (diesel)	59	–	4	–	30	617	126	400	3	13
– rail	2	–	–	–	1	26	5	15	–	1

(1) dust
(2) non-methane hydrocarbons

However, total energy consumption in the building sector (residential and tertiary) rose by over 20% between 1973 and 2005. The tertiary building stock has grown steeply (over 800 million m² heated in 2007), with an increase in the use of electricity for computers and air conditioning. For the residential sector, explanations include the increase in housing area per inhabitant (from 32 to 39 m² per person from 1990 to 2004), the rise in the number of residences (+14%, due to bigger population and increased numbers of single-parent families), and supplementary domestic appliances (dishwashers, freezers, etc.). The average accommodation surface area per inhabitant had already gone up by 25% from 1984 and 1992[373].

Around 300,000 new housing units are built each year, equally divided between houses and apartment blocks (with a trend towards individual housing; 61% houses from 1999 to 2004, which consume more than apartments). The average indoor temperature has gone up by 2°C in 25 years. The energy source split has changed considerably: consumption of oil (and coal) has been divided by 2 (by 8) since 1973, but gas and electricity consumption has been multiplied by 4. District heating and

[373]Eric Labouze, *Bâtir avec l'environnement, enjeux écologiques et initiatives industrielles*, Ed. de l'Entrepreneur, Paris, 1993

co-generation are fairly uncommon in France. Yet these systems mean that energy can be recovered with low CO_2 emissions (wood energy, geothermal, etc.). Overall, greenhouse gas emissions from buildings (residential, tertiary, etc.) rose from 90 to 110 million tonnes of CO_2 equivalent from 1990 to 2004[374].

Housing built prior to the first thermal regulations is responsible for 71% of residential emissions, which represents over 30 million houses and 2.2 billion m^2. New buildings thus represent around 1% per year of the total existing housing, but the rate of renewal is lower still: most new houses do not replace old ones that would have been destroyed, and they tend to correspond to an increase in the number of houses. The annual number of houses no longer in use is on average 83,000. The renewal rate is thus estimated at 0.2% per year.

Regulations now therefore concern existing housing. For buildings over $1000\,m^2$ built after 1 January 1948, and substantial retrofits (i.e. when building work over 2 years exceeds a quarter of a building's reference value[375]), a general performance threshold is required. In other cases (buildings of less than $1000\,m^2$, buildings dating from before 1948, more partial retrofits, in other words almost all retrofit operations), regulations apply per item: e.g. when a window is replaced, double glazing with reinforced insulation is required[376]. Thresholds are set on the thermal resistance of insulation, the efficiency of heating equipment, the consumption of ventilators, the loss from water tanks, the power installed for lighting, etc. These thresholds are not very demanding: e.g. a window's heat loss coefficient can reach $2.6\,W/(m^2.K)$, whereas using argon-filled double glazing and decent woodwork halves these losses so that the additional cost is thus covered in a few years. The risk is thus to nullify the improvement potential: installing inefficient double glazing makes it economically impossible over the thirty years of its lifespan (more if the frame is well cared for) to justify a second replacement with a better quality window.

Each inhabitant consumes on average 150 litres of water in his or her house and produces around 1 kg of waste per day (totalling 22 million tonnes per year). Each m^2 built represents from 1 to 2 tonnes of matter. In 2006, 420 million tonnes of aggregate were used in France, of which 20% for buildings and 80% for civil engineering[377]. Cement consumption totalled 24 million tonnes, of which 65% for the building sector[378]. Compared to the overall turnover of the industrial sector in France (500 billion euro in 1997), secondary matter represents 70 billion euro, of which 30 billion for building materials[379].

Worksite waste totals almost 50 million tonnes a year (cf. chapter 1). Waste is made up of 84% inert waste, 12% non-dangerous waste (e.g. metal, glass, plastic), and 4% dangerous waste (impregnated wood, used oil, plastic)[380]. One prevention

---

[374]Institut Français de l'Environnement, 4 pages No. 115, November–December 2006
[375]Estimated at a ratio of 1287 euro before tax per m^2 of net floor area for buildings used for housing (1100 euro for other buildings)
[376]Decree of 3 May 2007
[377]Waste forum, www.pole-risques.com
[378]Cahiers du CSTB No. 2773, December 1994
[379]Ministry of Industry, Department of industrial strategy, *L'industrie en quelques chiffres*, 1998
[380]Institut Français de l'Environnement, Waste from economic activity, Waste from buildings and public works, http://www.side.developpement-durable.gouv.fr/simclient/consultation/binaries/stream.asp?INSTANCE=EXPLOITATION&EIDMPA=IFD_FICJOINT_0000552

strategy to reduce these quantities is "deconstruction", which involves disassembling and sorting waste when the building's components are designed. In the 1980s, some countries (Netherlands, Denmark) put in place voluntary recycling policies, with the target of recycling 50% of demolition waste. In the early 1990s, the French average was estimated at 12%[381], but today the rate has exceeded 50%: 67% inert waste, 70% non-dangerous waste (e.g. 75% metals) and 77% dangerous waste. In France, production of recycled agglomerates doubled from 1987 to 1990, to reach 3 million tonnes, then 18 million tonnes in 2006 (i.e. 4%, compared to 30% in the Netherlands). Local programmes for managing building waste have been set up with several targets: eradicating wild dumping; setting up treatment site networks for recycling, recovery and storage; reducing landfill; participating in recycling efforts; using recycled materials to establish viable outlets and save on natural materials; and actively involving public works. The use of recycled aggregate in concrete requires increasing the cement content (by around 10%, depending on the mechanical property sought).

This chapter indicates several ways in which these various impacts can be reduced, using the appropriate techniques in design choices, in line with the methods described in the preceding chapter. An ideal solution rarely exists, but each technique brings advantages and disadvantages. For example, building with wood uses renewable resources, but its lightness may reduce thermal comfort. Insulating material from agriculture (e.g. straw, hemp, wool, wood fibre) uses simpler manufacturing processes, but is sometimes less efficient, either because of increased waste or a decrease in useful surface area. Electric heating may be interesting from a greenhouse gas point of view (using current French production and depending on the calculation method), but it generates more radioactive waste. Heating using wood saves fossil fuels but generates polluting emissions (VOCs and dust).

Depending on the local context, regional priorities can be based on different criteria, so that optimization varies. In a densely inhabited zone, for example, air quality is important because polluting emissions can significantly affect the health of a high number of people. On the other hand, manufacturing techniques can evolve, so that it is difficult to make a definitive conclusion on the choice of a product.

4.1 ENERGY EFFICIENCY AND RENEWABLE ENERGY

This aspect is essential, since flows (of energy, but also of water, cleaning products and waste) are cumulated over the building's usage lifetime. It is thus often preferable to make an initial investment to reduce these flows (e.g. high-quality heat insulation). The economic reasoning in terms of overall cost (including investment and management) can be used to show the cost-benefit of some technical options[382]. This approach is favourable to the environment thanks to its preventative effects.

---

[381] Eric Labouze, *Bâtir avec l'environnement, enjeux écologiques et initiatives industrielles*, Ed. de l'Entrepreneur, Paris, 1993

[382] Thierry Salomon and Stéphane Bedel, *La maison des négawatts, le guide malin de l'énergie chez soi*, Ed. Terre Vivante, Mens, 1999

In addition, several strategies take external costs (in particular impacts on the environment) into account when making decisions: "eco taxes" correspond to the "polluter pays" principle, while some tax reductions (e.g. heat insulation, renewable energy or replacing boilers) can act as incentives. These measures serve to underline the importance of environmental quality when choosing contracting parties, and it is worth preparing for them.

4.1.1 Heating

The objective here is to reduce heating loads, while ensuring a decent level of thermal comfort thanks to a suitably designed building envelope. This prevention strategy is based on bioclimatic design principles: ensure biological comfort by taking advantage of the climate. An approach like the passive house[383], can bring down heating consumption to less than $15 \, kWh/m^2/year$, for an inside temperature of 19°C, compared to an average of around $200 \, kWh/m^2/year$ (240 including Domestic hot water) in France. The average annual consumption for heating tertiary buildings is $138 \, kWh/m^2/year$[384] ($115 \, kWh/m^2/year$ for schools and 161 for offices). A value of $30 \, kWh/m^2/year$ (for needs and not consumption) has been set for the headquarters of ADEME (French Agency for the Environment and Energy Management) in Angers, and targets of under 15 are currently common. The COMFIE software was designed to support this kind of bioclimatic approach by making it easy to compare design alternatives.

The following actions need to be implemented in practice.

Reducing heat loss with good thermal insulation

Double glazing with reinforced thermal insulation has a Uv loss coefficient of around $1.8 \, W/(m^2.K)$, compared to 3.3 for standard double glazing. This performance is obtained using a thin layer of metal oxide (to the order of Angstrom, i.e. 10^{-10} m), which reduces radiation emissions from the warm pane towards the cold pane (low emissivity). The market penetration rate of these windows was only 20% in France in 2001 compared to 97% in Switzerland[385], but they are becoming increasingly common. Uv values of $1.1 \, W/(m^2.K)$ can be achieved by using rare gases (e.g. argon) in the intermediate space, and as little as 0.7 with low-emissivity triple glazing, even 0.6 with a low-conducting rare gas.

Argon-filled double glazing is widely used in Germany, but still too rare in France: the few additional euro spent on an air space are quickly paid back in energy savings. The advantage of triple glazing is debatable. In terms of the energy balance sheet, efficient double glazing (Uv of around $1.1 \, W/(m^2.K)$) is more beneficial than triple glazing on a south-facing façade because the solar factor of triple glazing is lower. On the other hand, triple glazing is more efficient on the other façades: this can be studied by dynamic simulation depending on the climate and the building's use. The economic balance sheet varies depending on local production of the windows (which had to be

---

[383] Annual heating requirement under $15 \, kWh/m^2$, air infiltration under 0.6 volume per hour with a pressure difference of 50 Pa (measured using the blower door test), cf. websites www.lamaisonpassive.fr and www.passiv.de

[384] *Les chiffres clés du bâtiment*, ADEME, Ed. 2006, www.ademe.fr

[385] Les cahiers techniques du Bâtiment No. 219, October 2001, pp. 95–99

imported until recently) and the price of energy (saving on cheaper windows brings the risk of long-term losses).

The overall heat loss coefficient of a bay window, U, depends on the Uv of the glazing and the performance of the frame. U can be as low as $0.7\,\text{W}/(\text{m}^2.\text{K})$ with the best triple glazing and insulating frame. Shutters or other mobile shades (e.g. outside blinds) can help bring down night-time heat loss by creating a radiative screen and reducing convective exchanges.

Efficient treatment of thermal bridges at production makes it possible to benefit from efforts made on the surfaces (flooring, walls and ceilings that give on to the outside or non-heated areas). External insulation can reduce thermal bridges along the perimeter of intermediary flooring (Figure 4.1).

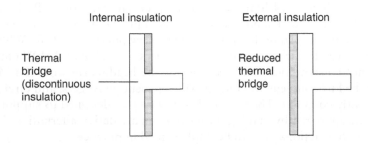

Figure 4.1 Incidence of insulation position on thermal bridges.

Different systems exist to reduce thermal bridges: reducing the contact surface between the floor slab and the wall (about 60% efficiency), or inserting an insulating block into the wall (about 30% efficiency, some breakers can be more efficient). To reduce thermal bridges at foundation level, two techniques exist: either using cellular concrete for the foundations (for individual houses for example[386]), or surrounding foundations with an insulator that can bear sufficient pressure (e.g. cellular glass or some polystyrenes). For glazing, placing the window frame in the insulating plane is recommended (cf. figure below).

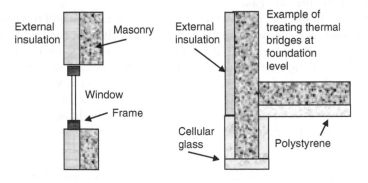

Figure 4.2 Thermal bridges at window and foundation levels.

---

[386]Cf. www.lamaisonpassivefrance.fr

On roof terraces, total insulation of the acroteria helps reduce thermal bridges.

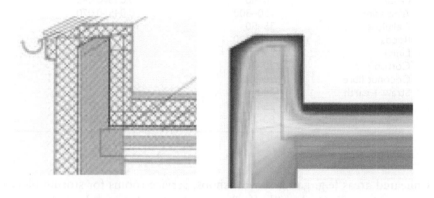

Figure 4.3 Thermal bridges treated at acroterium level[387].

External insulation also reduces variations in inside temperature since the thermal mass of the masonry is located on the premises side. Uniformly spreading insulation (e.g. using insulating bricks) gives an intermediate performance when it comes to heat. This kind of insulation is sometimes recommended for health reasons[388]. Fibre-based insulation can be separated from the premises by leak-tight partition walls, since some fibres can affect the respiratory system. Agricultural insulation solutions are an option, and their thermal resistance, and possibly acoustic performance should be ascertained. Table 4.2 gives several properties[389], whose evolution over time is not always known precisely.

As a comparison, the thermal conductivity of new glass wool is around 0.4 W/m/K. Vacuum insulation can achieve conductivity of under 0.01 W/m/K (values of 0.005 to 0.008 are given by the manufacturers). Passive houses have thus been built in Germany using 5 cm-thick insulation[390]. Thin reflective insulating material does not however manage this kind of performance, and a vacuum is necessary. Panels are partitioned so that the overall performance is maintained even when one of the compartments is pierced.

---

[387]Uli Neumann, European project TREES, Section 3 case study, 3.5 Nurnberg, http://direns.mines-paristech.fr/Sites/TREES/index.html, 2007
[388]Actes du colloque Santé et environnement: la brique pionnière, CSTB et Fédération des Tuiles et Briques, Septembre 2000
[389]Institute for agricultural and building research, Federal Agricultural Research Centre, Germany; and Friedrich Kur, *L'habitat écologique, quels matériaux choisir?*, Ed. Terre Vivante, 1999
[390]BINE Informationsdienst, Vakuumgedämmte fertigteile in der Baupraxis, Projektinfo 09/2007. www.bine.info

Table 4.2 Properties of some insulating materials.

Material	Density (kg/m³)	Heat conductivity (W/m/K)
Wool	10–40	0.035–0.045
Wood fibre	150–300	0.045–0.070
Cellulose	35–50	0.035–0.045
Reeds	100–200	0.04–0.07
Linen	25	0.036–0.038
Cotton	20–30	0.04
Coconut fibre	20–125	0.05
Straw + earth	340	0.09
Hemp[391]	25–140	0.04–0.07
Cork	80–150	0.04–0.05

Unheated areas (e.g. garages, workshops, service rooms for storing bicycles and sorting waste) can be used as "buffer" zones: located between heated areas and the outside (preferably on the north face), they reduce heat loss. A balcony or loggia can also be used as a buffer zone, closed by a glass window.

In well-insulated buildings, air tightness is an important parameter in the thermal balance, especially when dual-flow controlled mechanical ventilation with heat recovery is installed. This is because air that infiltrates is not pre-heated in the heat exchanger, which reduces the overall efficiency of the apparatus. Rendering a building airtight requires careful work (around frames, at floor slab-wall liaisons, plug sockets, etc.), cf. top of the figure below. Tests on the work site (in particular blower doors, infra-red thermography, cf. bottom of the figure below) can be used to check whether performance requirements of certain labels are met[392].

Protecting buildings from the wind has a fairly low impact on heat loss if walls are sealed and well insulated, but it can be useful to improve the comfort of outside areas or to protect a sunspace for example. The recommended distance between this protection and the house is 3 to 4 times the height of the protection so as to ensure a good compromise between protection against the wind and solar contributions[393].

Maximizing net solar gains

As we saw in chapter 3, the solar radiation that reaches the walls and roof of an individual house during the heating season in the northern half of France equals twelve times the house's heating load (60 times more for a passive house). In a typical construction, only 4% of this resource is recovered; the rest is reflected by the opaque surfaces and

---

[391]Hemp bark is called shive. Shredded, it can be used as to make lightweight concrete. The plant's stem is used to produce hemp wool.
[392]"Maison passive" label, cf. www.lamaisonpassivefrance.fr, and more recently the "Bâtiment basse consommation" label promoted by the association Effinergie
[393]Randall Thomas, *Environmental design*, Max Fordham & Partners, London, 1996

Figure 4.4 Measures to render buildings airtight.[394]

unused due to lack of sufficient thermal mass to store the heat. Solar gains therefore constitute important reserves.

Bioclimatic design should be incorporated right from the start of a project: it generally even concerns the choice of urban morphology. In a neighbourhood of Freiburg (Germany), buildings are placed in a row, with very open southern façades (cf. figures below). The rows of buildings are separated by green walkways (this neighbourhood is presented in chapter 5).

In fact, it is not necessary to organize the block plan so strictly: performances remain good with south-east and south-west orientations, so that it is possible to include less straight walkways or look for other ways of organizing the space.

Well-oriented glazed areas (preferably vertical planes between south-east, south and south-west) should be associated with protection against overheating in summer (roof overhangs, outside blinds, etc.) and high thermal mass (external insulation is recommended). The choice of window will depend on the climate, the orientation of the façade and the use of the building. The two main parameters are the heat loss coefficient and the solar factor (proportion of incident radiation received in the premises).

---

[394]Uli Neumann, European project TREES, Section 3 case study, 3.5 Nurnberg, http://direns.mines-paristech.fr/Sites/TREES/index.html, 2007

Figure 4.5 "Solar town" in Freiburg (Germany), source: www.quartier-vauban.de, now www.vauban.de.

Table 4.3 gives a few indicators for this choice. For example, on the south side of housing, it is best to fit argon-filled low-emissivity windows, possibly with low iron oxide content, to minimize loss and maximize gains (the southern façade is less exposed in the summer, especially if protected by an overhang). Triple glazing is suitable for the other façades, which receive less sun: double glazing's higher solar factor does not compensate for its higher losses. Mobile solar protection, which recovers solar contributions in winter while reducing overheating, is preferable to using windows with a low solar factor. However, to be efficient, this will require motorization or appropriate behaviour from the occupants. For this reason, "solar" windows are often used for offices, but solar gains are also blocked in winter time. Some blinds let light through but not heat (infra-red rays). Research is being done on thermo-chromic windows, which let through light but reflect heat above a certain temperature.

Table 4.3 Facts for choosing windows in a residental sector.

	Heat losses U (W/m²/K)	"Solar" window	Standard window	Window with low iron content
Solar factor (double glazing)		0.42	0.62	0.75
Standard double glazing	3.3	Non-heated premises only		
Low emissivity	1.7	Non-heated premises only		
Argon filled	1.3	West & east Offices	Mild climate	South
Low-emissivity argon-filled triple glazing	0.7 (0.6 with krypton)	West & east Offices	North, east & west	South cold climate

The above table can be used for small projects. For bigger buildings, a detailed thermal study is preferable. The same is true for tertiary buildings, where the choice of windows depends on the use of the different zones: temperature levels and internal gains can vary between offices, reception zones, meeting rooms, computer rooms, etc. In all cases, solar gains must be stored using suitable thermal mass.

Mobile solar protection is more efficient at combating overheating if it is located outside windows (e.g. rolling blinds, if possible with low emissivity and motor-driven from a central command). If these systems are closed at night, they can constitute a screen that reduces radiative exchanges and so heat loss during the heating season. Such equipment can be motor-driven, in which case cost and ergonomics need to be studied. Venetian blinds (inside or between two windows) reduce glare by redirecting light towards the ceiling. They are therefore suitable for work spaces (offices and schools, etc.). Detachable systems are preferable to "solar" (i.e. tinted) windows, which block rays even during heating periods, when solar gains would be useful.

A sunspace can be added to this set-up to preheat fresh air (e.g. double-door entrance that could receive a group of visitors). Dual-flow CMV (controlled mechanical ventilation) with energy recovery reduces the heating load linked to ventilation: the heat from the exhausted air is transferred into fresh air through a heat exchanger. The efficiency of this system can reach 80% with an efficient exchanger. Even greater efficiency is possible, but the pressure drop will be higher in the exchanger, which increases the consumption of the ventilators. Dual-flow CMV requires a second ventilator to blow in fresh air. To minimize electricity consumption and improve the overall energy balance, a high-efficiency ventilator is preferable, with a consumption of around 0.1 W for a flow of 1 m^3/h, compared to 0.25 for a standard ventilator. Extra-flat mini ducts make it easier to install double-flow systems in existing buildings. Parieto-dynamic systems recover some heat loss from buildings (and can even capture solar gains in the case of solar cladding): the fresh air circulates in the surfaces and recovers part of the heat loss from the envelope.

Specific solutions can take things further still, e.g. with "Trombe" walls, transparent insulation or active solar heating (air or water collectors, direct solar floors). Innovative windows provide insulation, solar gains in winter, solar protection in summer and produce hot water thanks to an association of collectors and mirrors inserted in double glazing containing rare gas[395].

Other illustrations are presented in chapter 5. Different types of components can be spread round building envelopes to ensure a significant share of the different energy requirements: e.g. thermal solar collectors for hot water, transparent insulation for heating, windows for lighting, photovoltaic cells to supply electricity, etc.

Choosing suitable equipment

Collecting and storing solar energy is not however sufficient to reduce energy consumption. To "transform" reduced heating needs into energy savings calls for equipment and appropriate settings. In particular, the zoning and location of thermostats (north rooms and south rooms), and thermostatic taps on radiators in south-facing rooms

---

[395] http://www.robinsun.com/

Direct solar floor (Photo. T. Letz)   *Passive solar gains*[396]

Figure 4.6   Solar heating for buildings.

to make better use of solar gains and improve comfort. If the setting is not sensitive enough, solar energy causes overheating instead of heating savings. In addition, settings must be adapted to intermittent occupation (week/weekend, day/night, etc.). Some systems recover the heat produced in certain zones (e.g. computer servers) and transfer it to other areas.

Settings must ensure the desired level of heat comfort in the different parts of the building. As a general rule (old people and young children usually require higher temperatures than young adults): living room and kitchen (19°C), bedrooms (their temperature can be set at a lower level, e.g. 17°C), bathroom (20°C). Based on simulations carried out on passive and standard houses (regulation RT 2005), an increase of 1°C in the house's temperature induces an increase of about 10% in heating demand.

The heating system must have high efficiency (at nominal operating conditions and at partial load). Oversized equipment generally leads to a drop in average annual efficiency. Unlike individual systems, collective central heating brings the possibility of having several cascading boilers. As a result, in mid season, only the necessary number of apparatus needs to be in operation, with efficiency much closer to that obtained at nominal operating conditions. On a boiler, a modulating burner can improve efficiency at partial loads.

An appropriate heating element can reduce the temperature of the heating circuit and therefore improve a boiler's performance. Low inertia systems (very thin radiators, air heating) are better at adapting to solar flow variations than more inert systems, such as floor heating. However, inertia activation integrating small capillary tubes (walls, floors, ceilings) can use small differences in temperature to heat or cool premises (e.g. night-time cooling and heat storage during the day in offices).

Putting water softener into the boiler circuit reduces the scale build-up on exchangers that leads to lower efficiency over time, but also brings the risk of corrosion. Descaling the installation can improve its efficiency. Insulating the pipes improves distribution and heat efficiency.

A heating system should also emit the least possible greenhouse gases, the least possible pollutants inside the premises (boilers with low emissions of NO_x and CO),

---

[396]Photo Saint Gobain Glass, Domus Solarhaus, Architect: W.P. Berndt

and generate the least possible long-term radioactive waste. For equal quantities of heat, natural gas is less polluting than other fuels: in particular, it emits 30% less CO_2 than fuel oil, and 40% less than coal[397]. French boilers release more pollutants than German equipment[398], e.g. over 70 mg/kWh of CO compared to 10 to 30 for gas boilers: catalytic oxidization reduces these emissions. Connecting to a heat system fed by an incineration plant, when possible, means that energy can be recovered from incinerated household waste, which is worthwhile from both an economical and ecological point of view.

The advantage of wood energy and biofuels is that they are renewable. However, it is expensive to fit an individual wood boiler with a device to filter dust and other pollutants released from combustion, and this solution is generally reserved to collective boilers. Wood heating using a district heating network is therefore an interesting solution. Open fireplaces give low efficiency (around 10%, up to 40% with very well designed inserts). Chimneys should be fitted with a closure to avoid opposite draught (and corresponding heat loss) when not in use. An open hearth releases toxic emissions (VOCs, dust, CO).

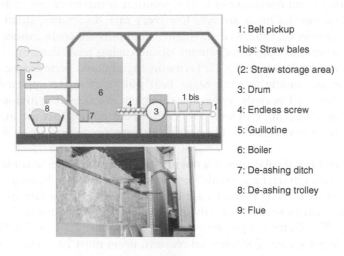

1: Belt pickup

1bis: Straw bales

(2: Straw storage area)

3: Drum

4: Endless screw

5: Guillotine

6: Boiler

7: De-ashing ditch

8: De-ashing trolley

9: Flue

Figure 4.7 Heating with straw.[399]

Electric heating results in a consumption peak in winter: power demand in France ranges from 31.6 to 89 GW[400]. As a result, power generators need to be installed that are underused in the summer (cf. figure below) and whose efficiency (25% to 30% including line losses) is much lower than that of boilers (from 75% to 95%).

[397] Eric Labouze, *Bâtir avec l'environnement, enjeux écologiques et initiatives industrielles*, Ed. de l'Entrepreneur, Paris, 1993

[398] Nadine Bonnier, Abdel Kader Lakel and Jacques Chandellier, *Caractérisation du niveau des émissions polluantes dans les fumées de petites installations de chauffage*, Les cahiers du CSTB, No. 3312, January-February 2001

[399] Photo CLER (Renewable energy liaison committe), Fact sheet: "*Valorisation de la paille pour le chauffage de serres horticoles*", ARENE Ile de France, Sept. 1996

[400] www.rte-france.com Technical results of the electricity sector in France, 2007

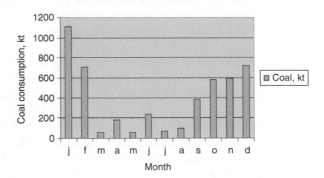

Figure 4.8 Seasonal variation of coal consumption in electric power plants.[401]

Electricity must also be imported, in particular in the winter (27.5 TWh in 2007, i.e. close to 6% of annual consumption). This solution is therefore not recommended in buildings unless heating loads are very low (very rare occupancy, small heated area, very efficient envelope), otherwise the higher maintenance costs cannot be justified. Heat pumps have the same disadvantage of increasing peak demand, but less so if their coefficient of performance (COP) remains high in cold periods (use of heat from the ground or an aquatic source). Some heat pumps require supplementary direct electricity during cold periods, either because they are sized for mid-season, or for technical reasons (freezing problems, etc.). Since heat pumps that use outside air as a cold source have a lower COP, it is preferable to use more temperate cold sources (e.g. exhaust air).

Seasonal storage of solar energy is not justified at house scale: sensible heat storage is only efficient on a much bigger scale (several hundred houses); hydrogen storage and electricity generation using fuel cells are currently very costly solutions. Geothermal energy can currently be used to heat 200,000 houses, thus economizing 230,000 tonnes of oil per year[402]. Water temperature must be sufficient (70°C to 75°C) to make it worth exploiting a spring. For financial reasons, users must be concentrated within a restricted perimeter around a pit.

4.1.2 Air conditioning

Passive cooling may be sufficient to maintain satisfactory thermal comfort levels in summer time, and avoid the noise and other nuisances generated by air conditioners. Along with selective masks (cf. paragraph 4.1.4), spectrally selective glazing (which transmits more light but less infra-red radiation) can ensure visual comfort while limiting overheating. However, to make the most of useful solar gains, this kind of glazing is more suitable for west façades (and possibly east) than for the south side, which receives significant sunshine in winter. Mobile protections can optimize overall performance, but this depends on the behaviour of the occupants, and on whether any

---

[401]EDF, technical operating results, 1997
[402]Eric Labouze, *Bâtir avec l'environnement, enjeux écologiques et initiatives industrielles*, Ed. de l'Entrepreneur, Paris, 1993

automatic controls function properly. The figure below shows the effect of an architectural mask (e.g. roof overhang above a window) on solar radiation depending on the season.

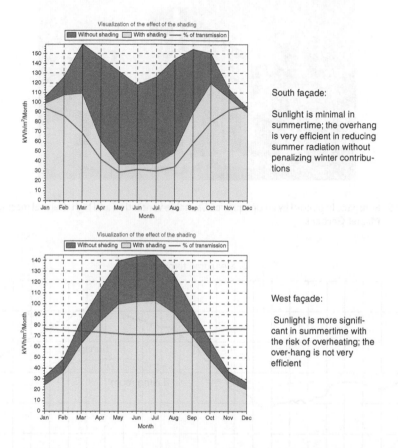

South façade:

Sunlight is minimal in summertime; the overhang is very efficient in reducing summer radiation without penalizing winter contributions

West façade:

Sunlight is more significant in summertime with the risk of overheating; the over-hang is not very efficient

Figure 4.9   Solar radiation with overhang (yellow area) and without overhang (green area): 1 m-wide overhang located 50 cm above a 1 m-high window (curve obtained by the PLEIADES_ COMFIE software).

Since daylighting is more visually efficient than artificial lighting, the latter should be used as little as possible. If thermal mass is spread in premises so that solar radiation crossing through the windows always reaches an inert component (e.g. floor, partition wall), then the energy is stored and the temperature variation will be smaller. Using the thermal mass of the ground to cool a building is possible in Mediterranean climates (fresh air is cooled down in underground pipes). The same system can be used in the winter to preheat fresh air. However, it is important not to damage the quality of the air, and these systems are not recommended if radon is present. Any condensation should be evacuated at a low point.

Night time ventilation can also help to cool down a building as a complement to thermal mass. These different techniques can even be used to keep the temperature of a sunspace at a reasonable level, as shown in the figure below.

Figure 4.10   Sunspace kept cool by an opaque roof, suitable ventilation and high thermal mass (Architect: Michel Gerber).

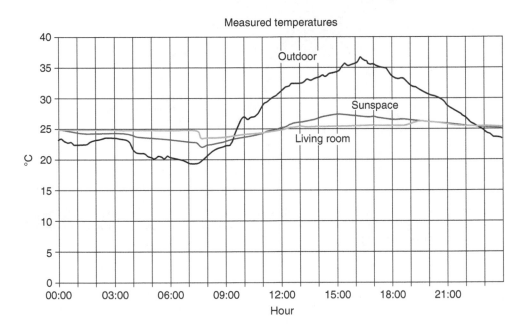

Figure 4.11   Temperature profile measured in a "bioclimatic" sunspace (source: Olivier Sidler).

Air conditioning systems should only be used in very specific cases: very high internal gains due to considerable occupancy (e.g. conference room) or machines, hot climates, noisy zones making natural ventilation impossible. A bioclimatic approach can be used to reduce the power and consumption of these systems. Desiccant or absorbing air conditioning systems can be solar powered.

4.1.3 Air renewal

To satisfy oxygen requirements, an individual needs to inhale around 0.03 litres per second, i.e. around 100 l/h[403]. In practice, respiration involves a much higher quantity, which varies depending on the activity from 6 litres per minute at rest to 15 while walking and up to 60 or 100 for an endurance race. Evacuating the CO_2 and vapour produced by breathing requires a high air renewal rate. The maximum CO_2 concentration is 0.5%. Maintaining CO_2 below this level, for people doing light work, requires a fresh air flow of 1.3 to 2.6 l/s (respectively 2.8 to 5.6 l/s for a maximum CO_2 concentration of 0.25%). The recommendation is 8 l/s for offices in the absence of smokers, and 10 l/s in the presence of moderate tobacco smoke[404].

Sufficient ventilation, generally 0.5 to 0.6 volumes per hour for housing, evacuates pollutants and humidity from inside the building. The decree of 24 March 1982 on housing ventilation fixes minimum ventilation flows based on the number of rooms. Buildings with high occupancy rates (tertiary, schools, etc.) require more air renewal[405], in which case it is a good idea to control ventilation either using a timer according to the occupancy, or a hygro-setting device: inflow openings and ventilator are controlled in line with the humidity rate in the room, which is linked to occupants' respiration. These devises can result in considerable energy savings without damaging air quality. However, air quality is more closely linked to the concentration of pollutants, in particular VOCs, than to the concentration of water. One possible occurrence is that the flow reduced by hygro-set ventilation becomes insufficient to evacuate the pollutants. It is therefore always preferable to choose surface coatings (walls and flooring) and furniture that release as few VOCs as possible (and especially formaldehyde) rather than having to remove pollution with increased ventilation that consumes more energy.

A sunspace or air collector (e.g. a perforated steel siding or a curtain wall) used to pre-heat fresh air brings both energy savings and greater comfort (by eliminating cold draughts). Dual-flow systems are probably the most efficient from a thermal point of view, but they require two ventilators for mechanical ventilation (natural dual-flow systems are still being researched[406]). Electricity consumption remains moderate if the system is well designed: in the case of single flow, the ventilator's power can be reduced to 15 W at a "normal" setting in a house, and go up to 60 W at a "high" setting. The average amount is 300 kWh/year[407], but this value can be reduced with more efficient ventilators that consume around 0.1 W to circulate 1 m³/h. Greater air flow in particular increases energy demand for heating.

---

[403] A.M. Mayo and J.P. Nolan, *Bioengineering and bioinstrumentation, Bioastronautics*, Macmillan, New York, 1964

[404] *Ventilation for acceptable air quality*, ASHRAE standard 62-1989, Atlanta, 1989

[405] Decree of 12 March 1978 on air renewal devices in buildings other than housing, Decree of 16 November 2000 on heating and ventilating premises

[406] This kind of system has been set up on a Building Research premises in London, as part of the BedZed project. However, results are not convincing: no detailed report has been provided on these operations.

[407] Olivier Sidler, *Maîtrise de la demande électrique, campagne de mesure par usage dans le secteur domestique*, final SAVE report, contract No. 4.1031/93.58, 1995

Natural ventilation can be an alternative to mechanical ventilation, but it needs to be carefully designed. Air inflows must be self-regulating to avoid draughts in line with the air temperature, wind speed and direction: they maintain a relatively constant (+/−20%) flow of air when the pressure difference varies from 1 to 25 Pa. Their size is for example 10 cm × 30 cm. Several systems exist: a steel blade is raised (aerodynamic system) or pushed (mechanical system) when the wind is excessive, thus reducing the opening: in electronic systems, the air speed is measured and the opening is adjusted by a motor. In this case, the slots can be manoeuvred automatically: e.g. to reduce flows at night and on weekends in offices during the winter, and increase the flow at night in summertime to cool down the building. For sites exposed to noise, acoustic air inlets are necessary. Hygro-adjustable slots can be set according to the rate of humidity inside the premises.

If an air inlet is placed close to a ceiling, the thin stream of air remains flattened against the ceiling thanks to the Koanda effect, thus avoiding the sensation of draughts. Systems also exist (that do not require energy) to automatically open windows: if the temperature of the premises reaches a certain level, a piston slips over a cylinder (filled with oil, which acts by change of state) and opens the window. The fusion-solidification can last from 30 minutes to 2 hours.

Air shafts extract used air via a flue effect (difference in temperature between hot exhaust air at the top of the building and cold air outside) and wind towers. Wind shields are 70 cm in diameter for a height of 50 metres. An internal cone protects against the rain (and birds), and a deflector reinforces the draw. The pressure drop must be around 2 Pa. A flap helps reduce night-time heat loss if ventilation is reduced.

In order to minimize heat loss in buildings, ventilation must be controlled. This involves making the building air-tight (e.g. plastic film can be used on the entire envelope of a wood-framed house). Woodwork permeability must be adapted to air renewal requirements. Air-tight letterboxes even exist to use on front doors. A double-door entrance could also be installed to limit air exchanges. To work efficiently, this entrance should be long enough so that there is time for the first door to close before the second one is opened.

4.1.4 Lighting and electricity consumption

Daylighting can be used to improve visual comfort while saving on energy, provided a device is fitted Cto counteract glare. Overhead lighting is effective from a visual point of view, but increases thermal discomfort in summer time. With some exceptions (e.g. lighting a stairwell), the best solutions strike a balance between heat and light, such as clerestories.

This kind of set-up lets winter sunlight through the glazed part. This radiation is reflected onto the very light surface of the opaque part and supplies indirect light in the premises, without the risk of glare. In summer time, the sun's high position and the roof overhang prevents radiation from reaching the glazed part. Other architectural approaches are possible to organize this type of "selective" overhead lighting. Light shelves can be used to avoid glare from vertical daylight and spread radiation round the room.

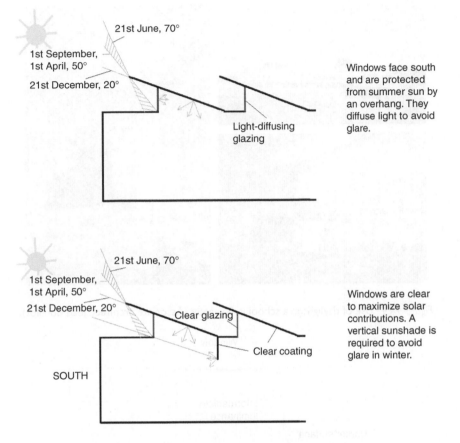

21st June, 70°

1st September,
1st April, 50°

21st December, 20°

Light-diffusing
glazing

Windows face south
and are protected
from summer sun by
an overhang. They
diffuse light to avoid
glare.

21st June, 70°

1st September,
1st April, 50°

21st December, 20°

Clear glazing

Clear coating

Windows are clear
to maximize solar
contributions. A
vertical sunshade is
required to avoid
glare in winter.

SOUTH

Figure 4.12 Design of a daylighting system on the roof.

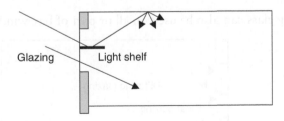

Glazing   Light shelf

Figure 4.13 Principle of a light shelf.

In the example in the figure below, bioclimatic design takes advantage of daylight and solar gains to heat a school thanks to wide, south-facing bay windows and additional light provided by a central atrium.

The bevelled design of the window embrasures increases the penetration of daylight and reduces the contrast by creating an intermediate luminance surface.

Figure 4.14 Use of daylight in a school in Baigneux (Associate architect: Jean Bouillot).

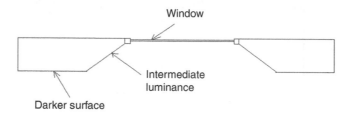

Figure 4.15 Design of an embrasure.

Light-diffusing glass can also be used on all or part of bay windows:

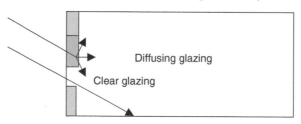

Figure 4.16 Combined use of a clear window and light diffuser.

Light-diffusing windows can be created using a capillary structure (usually plastic, polycarbonate or PMMA). In addition to their visual function, they improve thermal insulation: their heat loss coefficient is around $1.1\,\mathrm{W/(m^2.K)}$, i.e. three times lower than for standard double glazing. The solar factor is however a little lower.

Figure 4.17 Light-diffusing windows (transparent insulation) photo OKALUX.

The passage to daylight saving time helps reduce electricity demand for lighting, economizing from 0.7 to 1.2 TWh per year[408].

Artificial lighting should be designed to match usage, as well as the architecture: it can be organized into "zones", so that the occupants of a premises only turn on some lights, depending on the level of lighting in different parts of the premises according to their distance from the windows.

Compact fluorescent lamps consume eight times less energy than incandescence lamps, and correspond to standard visual comfort requirements (when very good colour rendition is not necessary). This type of equipment is suitable for continued use: turning them on and off too frequently (e.g. in a stairwell) reduces their lifespan. Electronic ballasts reduce the consumption of fluorescent tubes and extend their lifespan. Light-emitting diodes (LED) are also efficient: 2W bulbs exist that can be screwed into standard lamps, and desk lights fed by a USB cord plugged into a computer. The

---

[408]Ministry for the Economy, Industry and Employment, DGEMP (Department of Energy and Raw Materials, http://www.industrie.gouv.fr/energie/developp/econo/textes/se_heur.htm, 2008

light efficiency of the best LEDs is higher than that of compact fluorescent lamps, but this is not the general rule; the light beam may also be narrower.

By using energy-saving devices (lighting, refrigerators, dishwashers, washing machines, etc.), a family's annual electricity consumption can drop from 3200 to 2200 kWh/year for the same level of comfort [409].

Halogen lamps, which consume a lot of energy even with a dimmer switch, should not be recommended to occupants: the annual consumption of a single halogen lamp is around 250 kWh, i.e. half of the average total lighting consumption in a house. Efficient domestic appliances should be recommended (class A, cf. energy label to the right). Creating a space to dry washing avoids the use of a clothes dryer and the corresponding energy consumption.

Energy

NOTE

High performance

A
B
C
D
E
F
G

Low performance

Energy consumption kWh/year
based upon a calculation

350

Actual consumption depends on use conditions and site

Volume, refrigeration L   200
Volume, deep freezing L   100

Noise
(dB(A) per picowatt)   40

European Directive n°94/2/C.E.
on energy labelling of products

Machines left on standby consume a non-negligible quantity of energy: 33% more for televisions, double the consumption of a television for parabolic antenna modulators, and a 50% increase for decoders. Standby mode constitutes 97% of video recorder consumption.

Home automaton systems are sometimes used experimentally, e.g. to control lighting. Magnetic locking (chip card used as a key) automatically turns off some of the electricity when occupants are absent.

Connecting up houses and offices via a computer network can help reduce transport requirements. Alternatives exist (e.g. local offices) to pool user resources (or even knowledge) and avoid having to permanently maintain a room in a house for an intermittent activity.

New types of elevators are available on the market, e.g. counterweight lifts consume less than hydraulic lifts. Cabin lighting may be subject to operation.

Once electricity demand has been reduced, all or part of this energy can be supplied by a renewable energy system, e.g. photovoltaic[410]. The photovoltaic effect was discovered by Becquerel in 1839, and the first applications were made over a century later in the space domain. An electron (negative charge) that has received a certain quantity of luminous energy can free itself from the atom, thus creating a "hole" (positive charge). However, to obtain electric current, the electron must be prevented from immediately

---

[409]Olivier Sidler, *Maîtrise de la demande électrique, campagne de mesure par usage dans le secteur domestique*, final SAVE report, contract No. 4.1031/93.58, 1995
[410]Alain Ricaud, *Photopiles solaires*, Presses Polytechniques et Universitaires Romandes, 352p, 1997

PV modules for solar protection

Restaurant in Altenau, Germany (IEA Task 7[411]), Architect R. von Lamatsch Kämpfe

Fitted into the roof

Houses in Langedijk, Netherlands Photo Bear Architects

Semi-transparent cells in a sunspace roof

Environment education centre, Boxtel (Netherlands), Photo Bear Architects

Glazed balcony apron

Building in Aalborg (Denmark)

Figure 4.18 Photovoltaic installations integrated into buildings.

resuming its initial position. This is achieved by creating an electric field using positively "doped" layers (containing impurities chosen to produce a positive charge) and negatively doped layers. The free electrons can then be collected to produce a current.

PV modules can be integrated into the roof, the façade or act as a sunshade above bay windows (cf. figure above). Semi-transparent modules also exist, which can for example be used to cover an atrium or a sunspace. Flexible photovoltaic membranes can act to make the roof watertight[412]. A 3-hectare roof is under study in Carcassonne,

---

[411] International Energy Agency, database on photovoltaics: www.pvdatabase.org
[412] Cf. for example http://alwitra.de/wp-content/uploads/2013/05/evalon_fr.pdf

representing $1.4\,MW_p$[414], and another 2.5 ha roof is planned in Perpignan, a town that is aiming at a positive energy target.

The car park at Vaise Station in Lyon (778 parking places) is partially covered by $1180\,m^2$ of photovoltaic cells. The electricity produced is transformed into an alternative current by 55 inverters and sold to the grid, providing revenue of 21,000 euro a year, for a production of 150 MWh. The cost of the system is 630,000 euro (2006), i.e. less than 5% of the total cost of the construction. The system's overall efficiency is around 10% for crystalline silicon cells, meaning that around $10\,m^2$ of cells are required for a peak power (maximum power) of $1\,kW_p$, with annual supply of about $100\,kWh/m^2$. The direct current produced is transformed into an alternative current using an inverter, which can be used for standard appliances or fed into the grid.

On remote sites, electricity produced during the day can be stored in batteries. In general, the system's peak power P_c is set at 1.2 times the maximum power demand. The number of batteries can be evaluated to obtain a given capacity C (in Ampere.hour) depending on the voltage V (e.g. 12 V) by:

$$20P_p < C.V < 40P_p$$

For example, for a peak power of $3\,kW_p$ and a voltage of 12 V, around thirty 250 Ah batteries are needed. The photovoltaic system may be supplemented by a generator, or even a wind turbine.

Photo AWEA[413]

Photo WINDSIDE

Figure 4.19 Examples of domestic wind turbines.

---

[413] *American Wind Energy Association*
[414] Peak MW, i.e. the electric power supplied by $1000\,W/m^2$ sunlight and a cell temperature of 25°C

4.1.5 Domestic hot water

Water savers (cf. paragraph 4.2) reduce hot water consumption and thus energy. Mixer taps, which may be thermostatic, increase levels of comfort while reducing consumption.

Some domestic appliances (washing machines, dishwashers) can be linked up to the hot water circuit, saving an estimated 80% of the total electricity consumption of these machines[415]. Minimizing the distances between hot water production and supply brings down losses in the pipes (caused either by thermal conduction or the quantity of hot water stagnating in pipes and heated for no reason). Thermal insulation of pipes is important in premises that are not heated, air conditioned, or that risk being uncomfortable in summer time. A slower boiler regime is preferable when only hot water is produced in the summer. Individual metering of hot water can be an incentive to reduce consumption if accompanied by clear information for users.

A suitably sized solar water heater can save around 50% energy over a year (10% in winter and 90% in summer). In sunny regions of France (PACA, Corsica, etc.), profitability is achieved through tax credits. Thermosiphon and collector-storage water heaters are less expensive, but they are only suitable for temperate climates with no frost. Thermal solar power is a good choice for collective installations, where economies of scale are possible.

For individual solar water heaters, it is best to anticipate a collector surface of one m² per user in northern France (including Paris). This surface can be 25% less in central regions, and up to 40% less around the Mediterranean. The tank's volume should be 50 to 60 litres per user.

To qualify for aid, the materials used need to be certified, and collective installations must be designed and sized by an engineering firm to guarantee solar results.

In 2006, 1.2 million m² of collectors existed in France, totalling a power of 812 MW[416]. The area of collectors installed is 20 m² for 1000 inhabitants, compared to 350 m² in Austria and 300 in Greece. The long-term technical potential is estimated at 80 TWh, representing savings of around 7 million tonnes of oil equivalent per year (and nearly 10% of energy consumption in the form of heat).

The type of collector depends on the climate and the usage temperature. For cold climates, vacuum collectors operate with a low level of heat loss. For swimming pools, unglazed collectors are sufficient, either carpet-type (bundles of black plastic tubes) or metallic (more efficient but expensive). The collectors' area is equal to the pools' area in cold climates, and to half of that area in Mediterranean climates.

The figure below shows an example of a solar pool in Ile de France. The productivity of unglazed collectors is around 200 kWh/m²/year, compared to about 400 kWh/m²/year for glazed collectors, but they are much cheaper. The payback period for this type of installation is ten years.

---

[415] Olivier Sidler, *Maîtrise de la demande électrique, campagne de mesure par usage dans le secteur domestique*, final SAVE report, contract No. 4.1031/93.58, 1995

[416] *Le baromètre européen des énergies renouvelables*, Eurobserv'er, www.energies-renouvelables.org, 2007 edition

Thermosiphon

Separate elements

Figure 4.20 Solar water heaters.

4.2 WATER MANAGEMENT AND QUALITY

The diagram 4.22 below shows the average breakdown of domestic consumption.

Different types of water-saving equipment can be used to reduce consumption by up to 30% overall[417]: adjustable 3/6-litre flush systems or additional "Gustavsberg"

---

[417]Claude François and Bruno Hilaire, *Guide pour les économies d'eau*, Cahiers du CSTB No. 422, September 2001

Figure 4.21 Solar pool in Nemours.[418]

4-litre siphons (a flow accelerator ensures that the siphon remains efficient, but this system only works with a special toilet bowl), or dry toilets based on a compost system, time-delayed or infra-red controlled taps, flow restrictors, mechanical mixer taps for sinks and basins, thermostatic taps for baths and showers, and pressure reducers. Ceramic disks, which are harder than impurities (such as sand and lime scale) mean that taps remain leak-tight longer.

Sanitary equipment is designed to operate with a pressure of around 3 bars. Beyond that, wear and tear and scaling are accelerated, which increases consumption. A pressure reducer can bring significant savings. The figure 4.23 below shows a showerhead based on the Venturi principle. A ring placed on a spring works to reduce the diameter

[418]Photo CLER (Comité de Liaison Énergies Renouvelables), Fact sheet "Stade intercommunal de Nemours et de Saint-Pierre-lès-Nemours", ARENE Ile de France, September 1996

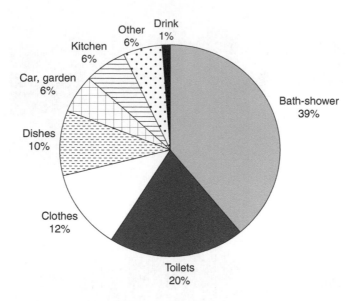

Figure 4.22 Breakdown of domestic use of potable water.[419]

of the passage according to the water pressure and so reduce the flow. To retain the same quality of use, the flow reduction is compensated by an increase in water speed using a cone. The reduction of the diameter of the passage leads to a higher speed and so a lower pressure (Venturi effect), drawing in air. This air, mixed with water, has a foaming effect that also increases the shower's efficiency and compensates for the reduced flow.

This kind of equipment can be complemented by rainwater collection for watering the garden or cleaning the car, in which case a reservoir must be installed (e.g. under a roof overhang). Some roof terraces with rainwater reservoirs contribute to regulating flow. Storing rainwater reduces the quantity of peak flow to evacuate, thus reducing municipal expenditure on drains.

Recovered greywater (or waste water) can cover around 30% of the needs of a domestic habitat. The cost price of water produced was estimated at 3 euro/m^3 in 1993, i.e. around double the average selling price in France at the same date[420]. It is more profitable to recover greywater when wastewater treatment is billed according to the quantities released into the drains (i.e. polluter pays principle). Each inhabitant releases 156 m^3 of wastewater per year. The corresponding treatment requires an average consumption of 46 kWh of electricity, 1 kg of concrete (for the pipe system), and transport by lorry of one tonne-km (sludge, etc.). The contribution to the resulting eutrophication is 3 kg of PO$_4^{3-}$ equivalent per inhabitant per year.

---

[419]T. Salomon and S. Bedel, *La maison des négawatts*, éd. Terre Vivante, Mens, July 1999
[420]Eric Labouze, *Bâtir avec l'environnement, enjeux écologiques et initiatives industrielles*, Ed. de l'Entrepreneur, Paris, 1993

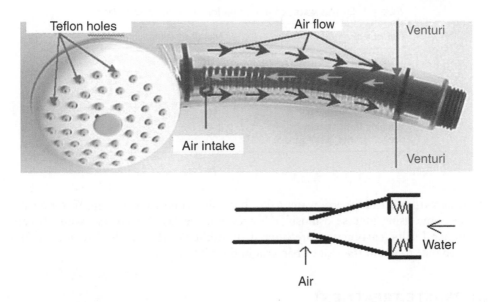

Figure 4.23 Showerhead with accelerated reduced flow using the Venturi effect (photo ÉCO-TECHNIQUES).

Around 21 kg of sludge is also produced, of which 50% to 60% is spread on farmland, 20% to 25% is put in landfill (banned from 2015), and 15% to 20% is incinerated. Urban sludge, which can be limed or composted, represents under 2% of the waste spread on farmland (94% is animal excrement). Manure spreading takes place from March to April and August to October, and is forbidden outside these periods. The sludge therefore needs to be stored for several months. Sludge can contain toxic components: heavy metals, organic compounds (polycyclic aromatic hydrocarbon and polychlorinated-biphenyl) and pathogenic germs (destroyed by heat treatment in water treatment plants). Using highly biodegradable detergents free from phosphates and toxic substances helps to limit the impacts of wastewater.

Dry toilets limit the release of substances that contribute to eutrophication, and produce fertilizer through compost. Another technique suitable for collective residences involves producing biogas in a fermentation reactor. This system can be complemented by vacuum aspiration to limit the use of water and its pollution.

As for energy, individual water meters are recommended, especially given that the issue of "heat theft" (heat transmission from one apartment to another) does not exist for water.

Leaks can be detected by measuring consumption at a time when it should be zero (e.g. at 4 a.m.). Water leaks can result in high consumption, cf. table below[421].

Many people choose to purify drinking water at home, a technique that avoids having to treat a large quantity of water in a sophisticated manner: an individual

---

[421] T. Salomon and S. Bedel, *La maison des négawatts*, éd. Terre Vivante, Mens, July 1999

Table 4.4 Surplus water consumption for different types of leak.

Type of leak	Annual consumption (m^3/year)
Oozing tap	1
Slight drip	5
Dripping tap	15
Slight leak in flush system	30
Thin stream from tap	90
Running flush system	250
Forgotten tap in the garden	500

consumes around 150 litres of water a day, but only drinks two of them. Water purified in the home using inverse osmosis is however more acidic than tap water because the mineral salts (carbonates) are filtered. In addition, this method requires careful maintenance to avoid the risk of bacteria in the filters.

4.3 WASTE TREATMENT

Waste sorting allows people to manage their waste more responsibly. Depending on local policies, adequate appliances need to be fitted into kitchens or other areas (e.g. garage or specific room, outside shelter), to store glass, paper, packaging (plastic, metal, etc.) and organic waste until collection. Other unsorted waste also requires a container.

A garden or balcony can be used to compost organic waste. This kind of waste represents 20% to 30% of all household waste. Kitchen waste presents a C/N ratio (ratio of the mass of carbon to the mass of nitrogen) of 20 to 30, which encourages rapid decomposition[422]. Compost can also be done on a neighbourhood scale. If a building lot contains a garden, then plants (lawn, shrubs, etc.) are a useful addition to the compost as long as there is an adequate split between grass, shrubs and organic waste. In gardens, some species require less energy for maintenance and generate less waste (e.g. a meadow can be mown two or three times a year).

4.4 BUILDING PRODUCTS

Given current building standards (2005 thermal regulation), the choice of materials has much less influence on the overall ecological balance than the flow management presented above. This is because the manufacture of products and building work are occasional, whereas flows of energy, water and waste cumulate throughout the building's lifespan. This is particularly so for existing buildings, which consume even more energy. To give an idea of scale, the energy consumption of the entire stock of houses

[422] *Manuel d'écologie urbaine et domestique*, Ed. Le Vent du Chemin, St Denis, 1992

is 70 million TEO per year, whereas all building materials industries together consume 3 million TEO. Those placing the orders should therefore be mainly motivated by products' functions and aesthetic advantages. However, when the design or retrofit of a building provides an occasion to increase usage performance, then the contribution made by the materials is significant in the overall environmental balance, as we shall see in the following chapter with the example of a passive house.

Choices need to be made between performance (which can result in reducing the quantities required), durability, non-toxicity, the use of recycled or renewable materials, or materials whose production and implementation creates little pollution. Based on the principle that no materials can meet all of these criteria at once, motivated choices need to be made. The use of certified products, when they exist, can make these choices easier.

Economizing on materials helps to preserve resources, as does the use of recycled materials (e.g. steel) or renewable matter (e.g. wood). One measure involves avoiding products judged to be useless (e.g. protecting steel components located inside a building can sometimes be avoided). Slab flooring or paving on outside areas can be avoided as much as possible, to save materials but also to render the ground more permeable. It is sometimes possible to do without a down pipe, by letting rainwater flow directly from the gutter into a tank from which it is collected. This however requires a site that is protected from the wind to avoid splashing the façade. Some project managers use laminated glass to avoid using woodwork around bay windows: the glass comes directly into contact with the masonry.

However, choosing materials is complex, since several elements are often contradictory. For example, wood is a renewable resource but usually has to be treated against damp and insects. Hardwoods are more durable, but only 1% of tropical forests are correctly replanted, which poses the problem of destroying resources and some ecosystems. A compromise must therefore be found, like using certified wood (e.g. the FSC label presented in chapter 1) that does not need protection.

Some types of wood (larch, acacia) come from replanted forests and are fairly durable. Species like spruce can be used inside the home (e.g. for stairs), while beech may be preferable for handrails because of its flexibility and hardness. Red cedar can also be used as a roof covering. A new wood treatment procedure, based on acetic acid derivatives and rapeseed or sunflower oil, can avoid the use of toxic substances. Impregnation is in two phases: a passage in an autoclave, then soaking in a bath at 100°C. According to the manufacturer, this process provides protection for ten years, can be used for all wood species and does not alter the colour or appearance of the wood. Protecting woodwork with aluminium can also avoid the use of polluting cleaning products, and the aluminium can be recycled afterwards.

Wood-frame constructions sometimes use pressed wood that releases formaldehyde (in low quantities, because the formaldehyde content of glue has dropped considerably), and the buildings' very low thermal mass makes it difficult to use solar gains (thermal mass must be included to store day-time energy during the night). Summer comfort is also penalized. One solution is to choose wood for the north part of the building, and a more massive structure for the southern zones. It is also possible to bring mass to a wood frame by integrating heavy partition walls, or even phase-changing materials (e.g. micro-capsules of paraffin mixed with plaster).

Increasing glazed areas on the façades can reduce the volume of masonry and thus of matter, but careful design is required to avoid an adverse effect on the energy balance and comfort levels (visual and thermal aspects).

Planted roof terraces retain rainwater and can integrate more easily into the landscape, although they require resistant structures. Hedges can be used to close off gardens and also play a role in reducing materials.

Waste from building work (demolition, worksites, renovation) is heavier than household waste. To facilitate taking the construction apart in the future, mixed materials should be avoided since they are difficult to separate at the end of their life cycle. Although it is technically possible to process composite products by separating their constituents, it is often difficult for economic reasons. Yet some composite materials last longer, which improves their environmental balance. Reusable tarpaulins can replace some packaging (e.g. plastic film).

Using pre-fabricated items often helps reduce worksite waste (in which case careful layout and specifications are recommended, a geometry corresponding to market standards being preferable to "made to measure"), but transport from the manufacturing unit can be longer.

Creating flexible areas can avoid waste in the development stage of a building, but because mobile or easy-to-dismantle partition walls are generally light and not leak tight, they may compromise acoustic and thermal comfort.

Concrete can be reused at the end of a building's life to make roads, but its manufacture requires quarrying. Crushed concrete can be used for walkways, car parks and some foundation parts, but it is still too fragile for use in construction (although in the Netherlands, most companies are committed to using 20% recycled aggregates in concrete used for construction). In France, recycled granulate is even more expensive to produce than natural granulate (on average 1.5 euro per tonne[423]). In the long term, recycling should become competitive because production costs of natural materials and the cost of access to class 3 landfills is set to rise significantly because of European regulations.

PVC window frames can also be recycled, but not profitably at the moment. However, pipes can contain a proportion of recycled plastic (in particular PVC and polyethylene). Recycled matter is also integrated into air-tight membranes and PVC floor coverings.

Raw earth is a mostly unprocessed material, but difficult to maintain. Several techniques exist[424]:

- Rammed earth (aggregate of earth and stones of diverse composition, taken from underneath farming land and compacted into a casing using a pounder);
- adobe (clay and sandy earth mixed with water and formed into blocks of different sizes);
- compressed earth (earth with more clay, compacted in presses to make rough bricks);

---

[423]Eric Labouze, *Bâtir avec l'environnement, enjeux écologiques et initiatives industrielles*, Ed. de l'Entrepreneur, Paris, 1993
[424]Friedrich Kur, *L'habitat écologique, quels matériaux choisir ?*, Ed. Terre Vivante, 1999

- cob (earth with a lot of clay, mixed with straw or other plant material and set on wooden trays);
- bauge (earth mixed with thin twigs and formed into balls piled on top of each other).

Stone and slate can be reused at the end of a building's life, but require an initial "environmental investment". Brick buildings also use raw materials (clay) and energy (1.8 more than concrete). Cellular bricks are more environmentally friendly, in particular when made from a process based on cereals proposed by an industrial from the agri-food industry, but apparently not yet used: a mixture of crushed corn and bran is dried to obtain a rigid foam. This foam (or e.g. expanded polystyrene) is mixed with clay. It disintegrates during the firing of the bricks, leaving air holes that form a porous, and thus insulating, material. Insulating bricks retain their insulating properties for much longer than most fibrous insulators (in which compression and humidity reduce performance), and do not release fibres into the atmosphere.

Cement manufacture requires less energy than in the past (25% reduction over the last 20 years[425]) thanks to a dry process method, energy recovery, and the use of ingredients that do not require firing (slag, fly ash, pozzolan, fillers). Thanks to the very high temperatures reached and the alkaline medium, fossil fuels can be partially substituted by waste (34% substitution fuels in 2002, compared to 48% in Switzerland[426]): 365,000 tonnes of animal meal and 36,000 tonnes of used tyres were employed in 2002 in kilns at cement works possessing the necessary equipment and authorization (e.g. old tyres covering 12% of energy needs at a factory owned by the Lafarge group, tar, used oil and paint waste can also be used).

The fact that the cement industry uses fuels emitting high levels of CO_2 (70% petroleum coke and 10% coal, releasing around 350 g CO_2 per kWh) and the decarbonatation process (under the effect of heat, limestone $CaCO_3$ is transformed into quicklime CaO and CO_2), mean that it emits some of the highest levels, at 14% of industry's total emissions (of which 65% due to carbonate removal). 33 production installations are subject to emissions quotas, totalling 14.2 million tonnes of CO_2 per year over the period 2005–2007, and 13.4 million tonnes per year from 2008–2012. The quotas for 2008–2012 are respectively 3.1 and 1.3 million tonnes for the lime industry (18 production facilities) and tiles and bricks (52 installations). Emissions are 525 g of CO_2 per kg of clinker produced. Cement only represents 10% to 15% of the composition of concrete (the other ingredients are water, sand and gravel), which explains the low energy content of this material.

Plasterboard can contain a by-product of smoke treatment (phosphogypsum). Scrap waste from manufacturing can also be recycled, and in future even used boards or demolition waste containing gypsum.

Materials from agriculture (hemp or cork insulation, etc.) can have a lower impact on manufacturing, especially if produced using organic farming methods (no fertilizer

---

[425] Eric Labouze, *Bâtir avec l'environnement, enjeux écologiques et initiatives industrielles*, Ed. de l'Entrepreneur, Paris, 1993

[426] Ministry of the economy, industry and employment, Industrial statistics and studies service (SESSI), *La protection de l'environnement*, www.industrie.gouv.fr/sessi/publications/etudes/mat/environnement.pdf

or pesticides). Care should also be taken to ensure that they do not generate additional impacts on the building's life cycle (cleaning products or increased waste, cf. table 4.2) and that they satisfy standard quality criteria (ACERMI certification or similar). Coconut fibre, straw and linen are used at the industrial stage to manufacture partition walls, insulation panels and cob.

Insulating materials based on polyolefin and polyester foam covered with aluminium can offer an interesting alternative to standard products, because they use less matter. They do not absorb water even in damp atmospheres, which means that they retain their thermal properties over time. On the other hand, their acoustic properties are mediocre and their thermal properties are controversial. They are therefore better suited to insulating roofs, eaves and ventilation spaces than façades exposed to noise.

Insulation made from recycled paper will become an interesting alternative when more paper is available. Currently, France is paper deficient, and has to import 500,000 tonnes of used paper and cardboard per year from the USA and Germany, incurring transportation impacts. Boron salt (borax and boric acid), which has a similar toxicity to table salt, can be used to protect recycled paper insulation from humidity, fungus and other living organisms, and also has a fire-proofing effect. Borate is not soluble in water, unless temperatures exceed 100°C, and it is sufficient to simply impregnate the paper (without the need for glue). Borate does not increase the toxicity of smoke in case of fire. Recycled paper insulation can be applied by insufflation (walls or roofs) or flock spraying, e.g. to create acoustic ceilings.

Some packaging waste can be repulped and agglomerated and then used to make agglomerated panels. Paper sludge, coal shales, coal dust, granulated blast furnace slag and fly ash can be used as light aggregates for concrete-cement. Recycling waste in building products should not, however, generate risks for occupants or builders. Examples of controversial projects include proposals to use animal meal in masonry and recycled radioactive waste in glass wool.

Labels exist for some types of product, like the NF-Environnement label for paint and varnish, CTB-P+ certification for wood treatment products, CTB B+ for the wood itself, and ACERMI for insulation. The table below gives a non-exhaustive list of these labels with the products concerned[427].

On the theme of floor coverings in particular, environmental quality needs to be balanced with the product's manufacturing and installation (glue, etc.), durability, influence on heating and cooling needs (an insulating floor covering may cancel out the thermal mass of a floor slab), and the amount of cleaning products used over its life span. A German association of carpet manufacturers created the GUT label for this type of product. The low thermal resistance of linoleum allows heavy floor slabs to play their thermal layer role, unlike carpets and floorboards. According to FDES (fact sheets on environmental sanitation), linoleum has a higher impact in its manufacturing phase than PVC flooring. However, these fact sheets do not account for phthalate emissions in the usage phase or dioxins emitted in case of incineration at end of life. The enamel on some tiles includes heavy metals. Some floor coverings contain significant proportions of recycled materials (e.g. tiling with 70% glass from crushed electric light bulbs, carpets containing recycled plastic and "natural" fibres, etc.).

---

[427] www.ecologie.gouv.fr/ecolabels/

Table 4.5   Different eco-labels.

Country	Label	Building products
Europe	Ecolabel[427]	Paint, varnish, floor coverings
Germany	Blue Angel[428]	Materials based on paper, glass or recycled plaster, floor coverings, wood products with low formaldehyde emissions, varnish with low VOC emissions, paint with low lead and chrome content
	GUT[429]	Carpets
Austria & Europe	NaturPlus[430]	Building products (masonry, dry construction, insulation, roofing, floors, paint, surfaces, etc.)
Northern Europe	Nordic ecolabel[431]	Floor coverings, boilers, panels paint, varnish, windows, chalets
Canada	Ecologo[432]	Insulation materials, floor coverings, materials based on recycled plastic, plaster tiles, coatings, wood, steel, proofing, paint
Japan	Ecomark[433]	Insulating materials free from CFCs & asbestos, cement with over 50% blast-furnace slag
Denmark	Indoor Label[434]	Climate Textile floor coverings and suspended ceilings (air quality)
France	NF-Environnement[435]	Paint and varnish
	CTB B+, P+[436]	Wood and treatment products
	ACERMI[437]	Thermal insulation
Certified equipment		
Germany	Blue Angel	Gas burners, condensation boilers, bathroom appliances (toilets & flush valves electronic showers), water flow regulators, low-noise construction machinery
Canada	Ecologo	Boilers and water tanks, heating, heat-recovery ventilators
France	Flamme[438]	verte Wood-burning stoves
Japan	Ecomark	Solar water heaters, solar hearing systems, valves & fittings

---

[427] www.eco-label.com/french/
[428] www.blauer-engel.de
[429] www.gut-ev.de
[430] www.natureplus.org
[431] www.svanen.nu
[432] www.ecologo.org/fr/
[433] www.ecomark.jp/english/index.html
[434] www.dsic.org/dsic.htm
[435] www.marque-nf.com
[436] www.fcba.fr
[437] www.acermi.com
[438] www.flammeverte.com

4.5 COMFORT AND HEALTH

4.5.1 Acoustic comfort

Technical solutions to protect a house or apartment from outside noise are provided in, e.g. acoustic regulations[439,440], and the Qualitel label[441]. Soundproofing between apartments encourages acceptation of grouped housing, which uses less energy. Protection against outside noise should at least conform to acoustic regulations (e.g. minimum insulation against road noise of 30 dB(A) if the façade is not on a listed road).

Bedrooms can also be protected against noise coming from other rooms. Minimum proofing from airborne noise – voices, television, music systems, etc., in comparison with pink noise[442] should be at least 41 dB(A) on emission and if possible 45 dB(A). The maximum transmission of impact noise – walking, dropping objects, etc. should not exceed 65 dB(A), and a lower value would be appreciated: 61 dB(A) is more in line with current user demand.

The maximum level of noise generated by systems is 45 dB(A) for an individual boiler if the kitchen opens out onto the living room, otherwise 35 dB(A), and 30 dB(A) for mechanical ventilation set at minimum flow.

Apart from the general envelope's soundproofing, the quality of the acoustic environment should also be considered. Floor and wall coverings should be chosen and combined to reduce sound reverberation: e.g. if the floor is covered with tiles, it may be necessary to lay an absorbent covering (e.g. cork) on one of the surfaces of the premises to avoid echoes. Furniture may of course also contribute to the sound quality.

4.5.2 Visual comfort

The priority should be to light rooms naturally (taking care to reduce the risk of glare), supplemented by efficient artificial lighting. Appropriate lighting levels and daylight factors are given in chapter 3 for different types of building (residential, industrial, tertiary, etc.).

4.5.3 Thermal comfort

As a general rule, in summertime, thermal comfort can be ensured by putting bioclimatic principles into practice, e.g. suitable solar protections (roof overhang, outside blinds), sufficient thermal mass, and appropriate ventilation. External thermal insulation can take advantage of the masonry's mass to reduce overheating (and recover solar contributions through windows during the heating season).

In Ile de France, for instance, the temperature inside a building should not exceed 26°C, even when the maximum daytime temperature is 30°C for several days in a row. When possible, night-time ventilation makes the building cooler.

---

[439] Mathias Meisser, Réglementation acoustique des bâtiments, Techniques de l'Ingénieur, Dossier n° C3365, August 2005
[440] www.ecologie.gouv.fr/Reglementation-acoustique-de-la.html
[441] www.qualitel.org
[442] White noise corresponds to an identical pressure level for all sound frequencies, whereas the emission spectrum of pink noise is representative of the noise emitted in the building

The most efficient protection (blinds or external rolling shutters) can also help to save energy by limiting loss through radiation (low-emissivity coating). Selective blinds block infrared rays but not light.

The cold wall effect is sometimes put forward to justify limiting glazed areas. However, this effect is low using contemporary high-insulation glazing, and the reason for reducing glazed areas is generally financially motivated.

Zéphyr system
(www.zefyrgroup.com)

Right, AWAX building in Athens
(Architect: Alexandros Tombazis)

Figure 4.24 Horizontal sun-break (overhang), right: vertical.

Individuals establish their comfort threshold to match their physiological state and lifestyle. Thus, one person might choose a carpet to walk barefoot in his or her house. Another person might choose tiles on a slab floor to increase the building's thermal mass, and thus save energy by making more use of solar gains and benefiting from natural cooling in the summer time. Taking environmental criteria into account can bring about behavioural changes without undermining quality of life.

4.5.4 Air quality and olfactory comfort

The main issues to resolve to improve indoor air quality are:

- protect against outside pollution;
- prevent the release of gases from decoration products and furnishings (paint, floor and wall coverings, furniture, etc.);
- expel foul air from living or work spaces, damp rooms (kitchen and bathroom), boiler room (combustion gas, odours), garage (exhaust fumes, petrol vapour) and other service areas;
- prevent proliferation of germs (legioniella pneumophila in particular).

Building sites should naturally be chosen bearing in mind sources of emission (industry, roads, etc.) and dominant winds. Air inlets should as far as possible be located away from any pollution sources (e.g. the street) and in the sunspace if this solution is chosen. If necessary (and in the case of dual-flow mechanical ventilation), a system can be installed for filtering outside air (dust and pollen, etc.), but this solution requires suitable maintenance. Pre-heated fresh air avoids the cold air sensation that encourages people to block up air vents. To this end, an efficient solution is to divide air inlets round the joinery work. Air outlets should be located in damp rooms. Controlled mechanical ventilation (CMV) with hygro adjustment or set to match occupancy associates air quality and energy savings. Good ventilation (of rooms and bedding) also helps reduce the allergy effect of dust mites and bring down humidity.

When radon is present in the ground, it is important to ensure that basements, under-floor spaces, walls, floorboards and pipe ways are leak tight, to ventilate the building's basement and under-floor space, and to ventilate the premises with a rate of air renewal in line with recommendations on radon in the air (cf. chapter 1). To limit radon contamination, a ventilated crawl space or a floor slab on a leak-tight ground surface is preferable to a basement, which communicates with the building.

Even when heating is electric, a flue should be planned so that occupants can change energy source in the future, e.g. for financial reasons, without undermining air quality. If an adjacent garage is foreseen, the passage from the house to the garage should be outside, with two doors close to each other sheltered by a canopy. This solution is preferable because an inside door brings the risk that exhaust fumes from cars could enter the house. To decorate furniture, the best option is products that do not release long-term irritating odours or harmful volatile compounds.

Some green plants can help purify the air. Ivy and sansevieria, for example, absorb benzene ($10\,\mu g$ per cm^2 of leaf), formaldehyde (2 to $3\,\mu g$ per cm^2 of leaf[443]) and trichloroethylene. Plant respiration gives off low quantities of CO_2, but the quantity is much lower than the amount absorbed during the day by photosynthesis. Plants also filter dust. They humidify the atmosphere, e.g. a ficus can release 10 to 20 grams of water per hour.

4.5.5 Other health aspects

Lead water pipes should be replaced to respect new drinking water standards. Brass taps release lead during the first months of use.

To avoid the risks of legionellosis in hot water facilities, storage and supply temperatures should respectively be above 60°C and 50°C. The tank and the whole circuit should regularly be heated up to 60°C (e.g. during the night) for one hour (at least 32 minutes), or to 70°C for at least one minute, to kill bacteria: at 50°C, the bacteria stop growing but survive[444].

Electric supply and circuits should be located away from beds when possible to reduce exposure to electromagnetic fields. Circuit breakers can be used to stop the circuit when no apparatus is being used.

---

[443] Eric Labouze, *Bâtir avec l'environnement, enjeux écologiques et initiatives industrielles*, Ed. de l'Entrepreneur, Paris, 1993
[444] CVC (Chauffage, Ventilation, Conditionnement d'air) No. 1/2, January/February 2001

Substances that generate toxic fumes in case of fire should be avoided as much as possible. A halon-free extinguisher reduces fire risks while preserving the ozone layer.

Access for the disabled should also be considered in the design, such as wide exits and space for turning round a wheelchair, cf. relevant documentation[445,446].

Storage for medicines and toxic cleaning products should be planned on a high level to reduce the risk of poisoning children.

Apart from human health, an environmental approach can also concern animals living close to housing, e.g. by creating areas suitable for nesting under roofs, hedges around the garden, ponds, etc.

4.6 CONCLUSIONS OF CHAPTER 4

Choosing techniques is not just a technical issue, it is a decision with economic, social and environmental consequences. One of the questions currently being asked is: Will the sustainable development approach significantly influence the choice of techniques in the building sector and will building standards evolve?

Techniques that were once marginal can become standard at another time, or in another country, and this standardization generally brings economic advantages (due to higher levels of production and distribution and easier implementation through habit, etc.): when double glazing became the standard in new buildings, its cost became similar to single glazing. The same will no doubt occur in the near future with argon-filled low-emissivity double glazing, which is currently the standard in Germany. Taking another example, renewable energy is better perceived now than it was 20 years ago, when it was widely considered as an ideological fad. With a view to taking a more rational approach to environmental aspects, several major trends in the evolution of techniques are summarized in the table below, which proposes a 4-stage strategy:

- Reduce needs through restraint (linked to behaviour) and efficiency (linked to energy-saving technologies)[447];
- Use renewable resources;
- Use supplementary solutions with lower impacts;
- Inform building users.

This strategy concerns energy, water and materials. As shown by the table below, it can be broken down into sub-sections (e.g. to reduce energy requirements: loss and recovery of solar contributions), then technical components (e.g. for loss: insulation, thermal bridges and ventilation). A more detailed version of this table could differentiate building items (e.g. insulation of walls, floors and roof, insulating glazing) then techniques (fibres, cellular materials), etc. The table does not aim to be exhaustive,

---

[445]Ministry for Ecology, Energy, Sustainable Development and Land Use Planning, Ministerial delegation for accessibility, documents and regulations,www2.equipement.gouv.fr/Accessibilite/rubriques/reglementations.htm
[446]AFNOR, Dossier thématique Accessibilité, www.afnor.org/accessibilite/reglementation.htm
[447]www.negawatt.org

but to show the variety of techniques implemented in a building and how they can complement each other in a consistent overall strategy.

The strategy of implementing low-impact techniques should not of course be to the detriment of users' comfort and health. Cost often hinders innovation, which justifies some degree of aid, in particular for demonstration projects. These aids can be viewed as a community investment to avoid certain environmental impacts: in other words, the external costs associated with some more polluting technologies (e.g. the cost linked to diseases generated by the emission of toxic substances) can be viewed as disguised subsidies for these technologies, if they are borne by the community. In addition, reasoning in terms of overall cost (investment + operation) often gives the advantage to low-impact technologies, which make energy and water savings.

Lastly, techniques alone cannot guarantee good performance: paragraph 3.8.4 shows how occupants' behaviour influences a building's performance. Restraint may be extended to moderate consumption of energy, water and materials, illustrated by the following examples:

- Turn off lights when leaving a room;
- Set reasonable temperatures; article 5 of the law of 29 October 1974, modified by the decrees of 3 December 1974, 5 August 1975 and 22 October 1979, is still in force; it sets the maximum authorized heating temperature at 19°C for housing, teaching premises, office buildings, premises receiving the public, etc. Unfortunately, this article is often not applied today. As for air conditioning, a setting 5 degrees lower than the outside temperature corresponds to the notion of "adaptive comfort" and is sufficient in most cases. According to article R. 131-29 of the French building code, "On premises in which a cooling system is installed, this system should only be turned on or kept running when the inside temperature of the premises exceeds 26°C" (except for specific cases, like swimming pools, skating rinks, museums, etc.).
- Reduce heating in case of absence (anti-frost setting), reduce temperature at night;
- Do not open windows for too long to air a room (5 minutes are sufficient to renew air in a room);
- Close shutters at night;
- Avoid using a car for short journeys;
- Take showers rather than baths;
- Repair water leaks as soon as they are detected;
- Choose plants that need little watering for the garden;
- Reduce consumption of packaging.

It is only when combined with a restrained approach that techniques like energy efficiency and renewable energy can be truly effective, and that ecodesign can genuinely improve the environmental quality of buildings.

Table 4.6 Summary of low-impact techniques.

Reduce needs		
	Energy	
	loss	insulation
		reduction of thermal bridges
		air tightness
		ventilation (control & pre-heating)
	solar gains	windows (heat and light)
		orientation
		thermal mass
	cooling	solar protection
		thermal mass
		night ventilation
	Water low-flow	taps
		showerheads
		pressure reducers
		low-flow toilets
	Materials	
	Reduced material	lighter walls if mass unnecessary
		precise quantities
Use renewable resources		
	Energy	
	thermal solar	hot water production
		active heating
		passive heating
		and cooling systems
	photovoltaic solar	electricity production
	biomass	heating
	Water	toilets
	rain water	watering
	Materials	
	structure	wood (e.g. FSC label)
	insulation	agricultural products (e.g. straw, hemp, wool, cork)
Limit impacts		
	Energy	
	fuel	wood, gas
		geothermal
		district heating
	equipment	efficient boiler
		cogeneration
	lighting	energy-saving
	domestic appliances	class A appliances
	Water	toilets
	wastewater	watering
	Materials	
	recycling	insulation (paper, clothing)
		metals
		concrete (aggregates)
	re-use	woodwork
	low-impact	cf. Ecoinvent or Inies data

(*continued*)

Table 4.6 Continued.

Inform users		
	Management	
	energy	thermostat settings
		ventilation settings
		insulating shutters or blinds
		less transportation
	water	behaviour
		metering
		leak detection
	waste	sorting
	Maintenance	
	energy	heating equipment
	water	leak detection
	wood, steel, etc.	paint, preservation, etc.

Chapter 5

Applications and constructions

This chapter presents case studies that show how eco-design tools have been applied, describes achievements implementing innovative technologies and their performance, from house scale to neighbourhood scale and including retrofit, then suggests prospects for progression.

5.1 DETACHED HOUSES

Can a detached house fit in with the concept of sustainable development? In comparison to apartment buildings, houses lose more heat because of their bigger outside surface (low compactness); their morphology incurs higher consumption of materials, and contributes to urban sprawl. Positive arguments include the fact that solar collection is often easier and that houses with gardens may reduce some leisure journeys, make compost easier and even local waste water treatment. This kind of comparison could include a life cycle assessment, taking into account local context (distances from work place, shops, waste treatment facilities, etc.).

In any case, individual houses can provide an opportunity to test technical innovations, since experiments are cheaper on a smaller scale, and thus move the whole sector forward. The examples presented here show how approaches have developed over time in France, and how important it is to take an overall approach to reduce the environmental impact of buildings.

5.1.1 Aurore housing estate (Ardennes)

Following on from the solar solutions created in the 1970s and 1980s as a response to the oil crises, the objective of this operation was to perfect passive solar heating concepts ("Trombe walls"[449]) and active solutions (air collectors) using new German technology: transparent insulation.

The experiment involved a group of six wood-framed houses located in Mouzon in the Ardennes region, northeast France. Transparent insulation was used for the first time in France in this operation carried out by the social housing company ESPACE HABITAT in 1992.

---

[449] Cf. http://en.wikipedia.org/wiki/Trombe_wall

The aim of transparent insulation is to make thermal solar collection more efficient by maximizing the greenhouse effect. The materials developed according to this
principle, based on German research[450], have a solar factor comparable to that of
double-glazing, and the heat loss is three times lower. It was therefore worth studying the energy benefits of this technology in buildings. The first experimental solar
walls, for instance in houses in Odeillo (Pyrenees) in 1974, were single-glazed, which
reduced their performance in comparison with opaque insulation because of significant
losses.

The AURORE solar housing estate, designed by the architect Jacques Michel[451],
constitutes a particularly interesting experimental platform because active systems (air
collectors) and passive systems (Trombe walls) can be compared.

Figure 5.1   Solar housing estate in Mouzon (Ardennes).

The houses are south facing, and dissymmetrical to the two streets that border
the estate: French windows and solar walls can be seen from the street to the south,
whereas the street to the north looks onto opaque façades. The houses are identical to
keep costs down, with appropriate design (French doors in living rooms facing south,
solid doors in halls facing north).

Using on-site experimentation, performances estimated by calculation or measured in the laboratory can be verified, taking into account the effects of occupants'
behaviour. Measurements of productivity are only a small part of the evaluation,
which also focuses on comfort, implementation issues, costs, durability, aesthetics,
etc. Results were monitored for a whole year, covering both the heating and summer
seasons.

---

[450] A. Goetzberger et al., *Special issue Transparent insulation*, Solar Energy, Vol. 49, No. 5,
pp. 331–449, November 1992
[451] J. Michel, patent ANVAR TROMBE MICHEL BF 7123778 (France 29/06/1971) and add-on:
"Stockage thermique" MICHEL DIAMANT DURAFOUR 75-106-13

5.1.1.1   System operations

- Passive system (houses 3, 3bis, 7 & 9)

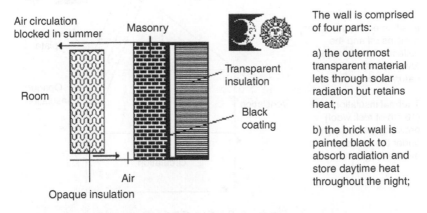

The wall is comprised of four parts:

a) the outermost transparent material lets through solar radiation but retains heat;

b) the brick wall is painted black to absorb radiation and store daytime heat throughout the night;

c) when air comes into contact with the bricks, it heats, rises and circulates towards the living area (air vents are open in winter in fine weather);

d) a partition wall insulated by 10 cm of rock wool protects from summer heat or winter cold in bad weather (when air vents are closed).

Figure 5.2 Principle of the passive system.

- "Active" system (houses 4 & 5)

Air is heated when it comes into contact with a black surface, the absorber, located underneath the transparent material that forms the roof (cf. figure below). When the absorber's temperature exceeds that of the house, a ventilator starts up and blows the hot air into the house. A thermostat stops the system if it gets too hot inside the house.

In all of the houses, the air is renewed by controlled mechanical ventilation (CMV), i.e. continuous flow ventilators.

In both set-ups, two types of transparent component are compared:

- a capillary polycarbonate structure, encapsulated between two panes and manu-factured by the company OKALUX[452] (houses 3, 5 & 7);
- an alveolar sheet of extruded, triple-wall polycarbonate, manufactured by CELAIR (houses 3 bis, 4 & 9).

These two materials improve the Trombe walls' productivity compared to single glazing. The corresponding physical properties are given in the table below; the solar transmission values were measured by CSTB on the OPTORA platform[453] in Grenoble.

---

[452] www.okalux.de
[453] www.cstb.fr

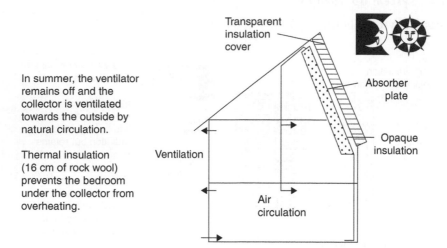

In summer, the ventilator remains off and the collector is ventilated towards the outside by natural circulation.

Thermal insulation (16 cm of rock wool) prevents the bedroom under the collector from overheating.

Figure 5.3: Principle of the active system

Figure 5.3 Principle of the active system.

Table 5.1 Properties of the two transparent surfaces.

Material	Heat loss $W/(m^2.K)$	Solar factor at normal incidence
10 cm of capillary structure encapsulated between two panes	0.8	0.72
triple-wall 16 mm polycarbonate sheet	2.4	0.70

Table 5.2 Cost of different alternatives (French Francs, 1992).

Configuration (2×50 habitable m^2)	Total cost	Supp. solar cost	Cost per habitable m^2
Passive, polycarbonate (13 m²)	479,000	29,010	4,790
Passive, 10 cm OKALUX (13 m²)	523,000	73,084	5,230
Active, polycarbonate (17 m²)	503,000	53,275	5,030
Active, 5 cm OKALUX (17 m²)	546,000	96,045	5,460

The cost of the houses and the supplementary cost of the different systems are summed up in table 5.2, in 1992 French francs, including tax. The cost of linking up to roads and networks and the price of the terrain are not included.

5.1.1.2 Assistance provided on the project's design

The thermal calculations that we did for the architect, using COMFIE software, produced an initial evaluation of the energy performance. The simulation tool was not used to size the systems: the architectural design was already very advanced, and it was the budget in particular that determined the collection surface area. However, numerous sensitivity analyses were carried out (cf. figure below) to assist in the choice of building parameters, like thermal storage materials, masonry thickness, absorbing covering, transparent insulation thickness, etc.

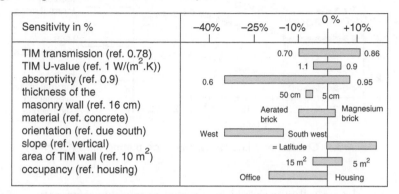

Figure 5.4 Sensitivity analyses of solar wall productivity.

These results imply that the covering of the absorbent surface plays a significant role. From an aesthetic point of view, the transparent insulation masks the dark surface that critics of the Trombe wall have disparaged. Productivity is linked to the use of the building: it is lower for offices, which are only heated during the day and benefit from high internal gains. The theoretical research helped better define the experimental project that was achieved.

Simulations were carried out for a typical year corresponding to the climate in the town of Nancy. The table below gives results for both systems and both types of material. The solar fraction, taking into account all of the useful gains through glazed and opaque surfaces, ranges from 30% to 45% depending on the system. In comparison, doubling the thickness of the opaque insulation from 10 cm to 20 cm on the walls would gain 13 kWh/m²/year, i.e. around 10 times less. The transparent insulation surfaces are 13 m² for passive systems and 17 m² for active systems. Houses are made up of two 50 m² levels. The calculations take the temperature in the houses to be a constant 19°C.

Table 5.3 Comparative results for the different systems (simulation with comfie).

Configuration habitable surface area 2 × 50 m²	Heating load (kWh/y)	Load per m² (kWh/m²/yr)	Solar fraction (%)
Trombe wall, polycarbonate	7696	77	31
Trombe wall, 10 cm OKALUX	6965	70	35
Active system, polycarbonate	6030	60	40
Active system, 5 cm OKALUX	5339	53	44

The control of active systems was studied, and the ventilation flow was reduced to diminish nuisances (noise, dust, draughts). The thermal calculations were also used to estimate overheating risks, both in terms of comfort and the durability of materials:

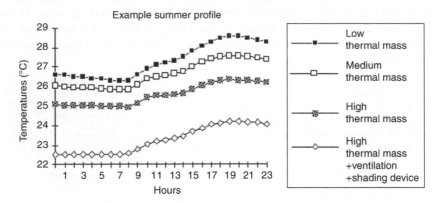

Figure 5.5 Comfort on a hot day depending on thermal mass (COMFIE).

stagnation of the air collector should be avoided because its temperature could reach 130°C, which would cause the polycarbonate to deform.

Simulations showed a risk of overheating in summer. However, this risk is due to the very low mass of the houses, and not the solar heating: the collecting components are separate from the habitable rooms and the air circulation is stopped during the summer. During an average summer, careful management can diminish the lack of comfort. Figure 5.5 shows how the comfort would improve by increasing the thermal mass of the houses (i.e. replacing the wooden frame with heavy masonry).

5.1.1.3 Description of the experimental installation

The external temperature and vertical southern and horizontal solar radiation are measured every quarter of an hour. The temperature of the house is measured on the ground floor and in the three upstairs bedrooms. The temperature is also measured in the collectors (on three levels) and in the Trombe walls, in the internal and external air spaces. The operation of the ventilators is noted, along with the total consumption of electricity and heating.

All of the measurements are linked to data loggers, with one logger for every two houses. These data loggers are linked to a telephone network accessible to different research centres. Every night, the loggers are called to transfer the data onto a computer. This experimental device was set up with the technical assistance of CSTB.

5.1.1.4 Monitoring results

a) Results of the summer 1992 monitoring

Two aspects were studied:

- Thermal comfort in the houses, on the ground floor and in the three bedrooms on the first floor, with different orientations (south-east south-west, north-east);
- Respect of acceptable temperature limits by the materials at sensor level.

The simulations carried out revealed overheating risks in the summer. The analysis showed that overheating is not due to solar heating, but to the absence of thermal mass. The simulation on the reference house, not equipped with solar heating, also showed high temperatures (solar heating adds a maximum 1°C in the bedroom located underneath the collector).

The results of the measurements confirmed these forecasts. The summer of 1992 included a very hot period of four days around 8 August, when the maximum outside temperature reached 38°C. Apart from this exceptional period, temperatures remained reasonable: during moderately hot days (maximum outside temperature between 25°C and 30°C), temperatures inside the houses did not exceed 28°C, which is within the comfort zone. The hottest bedroom was not the one located under the collector, but the one under the skylight, which is not equipped with solar protection (the difference between the two rooms was very slight). The bedroom located to the north had a temperature of at most around 1°C less than the other rooms.

The maximum temperature admissible for polycarbonate is 130°C. Temperatures above 100°C and high day-night variations could possibly affect its durability, although this is still not very well known. Risks must therefore be evaluated at this level, bearing in mind that polycarbonate sheets have been used in sunspaces for about twenty years.

The temperature was measured at the surface of the black absorber. The transparent coverings are in fact not as hot: their temperature is somewhere between this maximum and the outside temperature. In addition, the probes are exposed to solar radiation and therefore indicate a slightly overestimated temperature. The maximum temperatures were reached during several "heat wave" days, i.e. a short time. They are presented in the table below.

Table 5.4 Maximum temperatures in the collectors.

	Active system	Passive system
alveolar polycarbonate	105	65
transparent isolation	95	80

The performance of the materials used can therefore be assured, as long as the active collectors are correctly ventilated. The maximum temperature was obtained with alveolar polycarbonate (around 10°C more than with transparent insulation). The explanation for this is that the collector on house 4 (active system, alveolar polycarbonate) was less well ventilated because the air inlet meter located in the entrance hall was closed (even in these conditions, the collector was sufficiently ventilated thanks to vents in the gable). The temperature in the transparent insulation solar walls was above 15 K compared to a standard alveolar polycarbonate covering. This can be explained by the thermal insulation, by which heat is more efficiently stored in the masonry, in particular during the night.

b) Results during the autumn

We made a detailed study of the period from 25 October to 1 November, which comprised 3 cloudy days followed by 4 sunny days and one gloomy day. The period is therefore a good representation of mid-season weather. The outside temperatures ranged from 5°C (night-time minimum) to 12°C (day-time maximum) at the start of the period and dropped to −2°C to +3°C on the last day. The data are shown in the figure below.

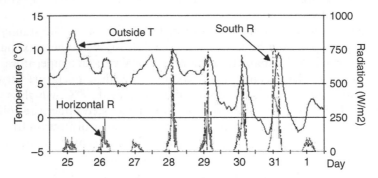

Figure 5.6　Typical weather data for an autumn period (25/10/92–01/11/92).

During fine weather, the air collectors work well as a heating system: their temperature can rise to 43°C, and up to 52°C with transparent insulation (figure below). They do not work in bad weather, when their temperature does not exceed 25°C (it can drop to 5°C at night, there was a frost on the last night). The temperature inside the houses ranges from 15°C to 25°C, depending on the sun and the thermostat set point chosen by the occupants. The setting cuts off the ventilator if the house's temperature exceeds 24°C, thus avoiding mid-season overheating.

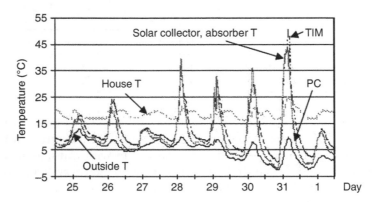

Figure 5.7　Temperatures in active systems, 25/10/92–01/11/92.

Solar heat is not as easy to measure in passive systems, since the wall stores day gains during the night. As a result, there is not a significant flow of hot air, as in the preceding case. However, the wall remains warm day and night. The reduction in heat loss also results in heating savings.

Solar walls receive radiation that passes through transparent elements. Their outside surface heats up, with temperatures reaching 47°C and even 64°C with transparent insulation (figure below). This heat is stored in the bricks of the walls, so that the temperature remains warm on the inside, ranging from 15°C to 25°C. When this

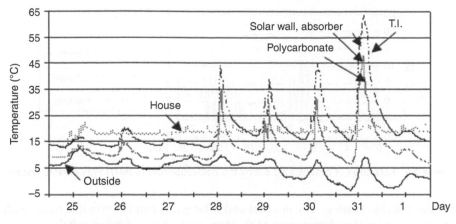

Figure 5.8 Temperatures in passive systems, 25/10/92–01/11/92.

temperature exceeds that of the house, the air starts to circulate (provided the air vents are open).

Because storage in the bricks significantly attenuates the amplitude, the variation in temperature is much smaller than in active houses. It is therefore the occupant who sets the temperature in the house using the thermostat. A thrifty occupant will reduce heating during the night and heat bedrooms less, while another might constantly keep the heat between 21°C and 22°C. This explains the significant consumption differences observed between houses.

c) Results during the coldest winder period

We analyzed the week from 31 December to 7 January, during which the temperature dropped to −12°C. The first four days were fine, the next two were variable, and the last two were very cloudy but warmer (5° to 7°C).

Active system

During the first four days, the temperature in the collectors rose as high as 30–35°C, then to 25°C on the fifth day, 20°C on the sixth, and remained at 10°C during the two cloudy days. The temperature inside the houses varied depending on the setting chosen by the occupants. On the finest days, the ventilator operated from midday to 6 p.m. During this period, the heating consumption dropped significantly: around one third in comparison to the night. On the cloudy days, the ventilator did not operate because the collector temperature was too low.

Passive system

On the fine days, the temperature between the brick wall and the transparent component (on the black painted surface) rose as high as 55°C with the capillary structure, and up to 40°C with the alveolar sheet. On cloudy days, it dropped to 12°C or 8°C. The temperature fluctuated a lot less on the inside, due to the thermal mass of the brick wall. It rarely rose above the temperature of the living room, but remained pleasant: the wall reduces heat losses, even though, during very cold periods, it does

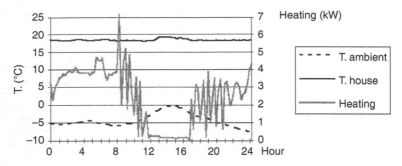

Figure 5.9 Reduction of heating load during the coldest period (02/01/1993), passive system.

not really provide heating. The internal and solar gains make it possible to do without supplementary heating during part of the day, as shown in the figure below.

d) Overall energy balance

The heating consumption readings give the following gross values, shown in the table below together with the estimates.

Table 5.5 Consumption forecasts and measurements over a heating period.

House	Consumption measured (kWh)	Consumption calculated (19°C)	Average actual temperature	Actual degree hour
3 (passive, TIM)	3560(*)	7000	16.32	52,496
3 bis (passive, PC)	8218	7700	18.97	65,787
4 (active, PC)	3050(*)	6000	18.51	(1)
5 (active, TIM)	4685	5300	18.64	(1)
7 (passive, TIM)	7037	7000	18.75	65,255
9 (passive, PC)	5051	7700	16.74	55,107

(*) The CMV was stopped in this house, CMV would add 3500 kWh
(1) A 6-week period is missing due to technical problems

The differences between the calculations and measurements are mainly the result of occupants' behaviour: the simulations considered a constant temperature of 19°C (i.e. 66,540 degree hours), while some occupants turned off the heating during working hours and lowered it at night, others heated above 19°C. In addition, some occupants stopped the CMV, although we considered a constant ventilation of 0.6 volumes per hour in our calculation. The corresponding heating requirements are around 3500 kWh. Closing blinds, internal gains (e.g. halogen lamps) and manipulation of vents on the Trombe walls can also explain some differences between forecasts and measurements.

To compare the systems, it is therefore necessary to use an identical occupancy scenario in all the houses to eliminate the effect of occupants' behaviour. This is why the simulation is useful, provided of course that some parameters (especially thermal bridges) are reset according to the measurements. As the figure below shows, once the

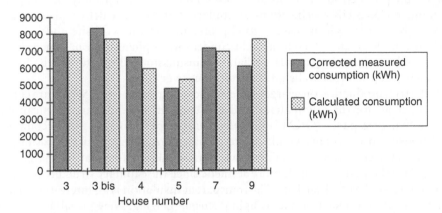

Figure 5.10 Comparison between consumptions calculated and measurements corrected according to occupancy scenario 3, 3bis, 7, 9: passive systems; 4 & 5: active systems.

corrections have been made to the degree hours and CMV, the measurements are fairly close to the forecast calculations.

According to the social housing company, the average heating consumption in the region for this type of house (100 m² of habitable surface area) was around 10,000 to 11,000 kWh (in 1992). The heating savings are therefore about 3000 to 5000 kWh depending on the system. To facilitate continuous measurements, a backup electric heater was used, which increased the heating expense. Some occupants then used oil stoves and their oil consumptions were metered. Solar savings appear to be viewed more favourably by inhabitants with, e.g. a backup gas heater, which is cheaper to operate. To explain the process to inhabitants, a brochure was published with recommendations and tips.

5.1.1.5 Conclusions of the experimental monitoring

Solar heating therefore resulted in significant savings without a drop in comfort levels. Transparent insulation, compared to standard double glazing, doubles the productivity of Trombe walls and increases that of air collectors by 25%. Total heating savings are around 20% for the passive system and 40% for the active system. Consumption ranges from 50 kWh/year/m² for the most efficient system (active system with transparent insulation) to 80 kWh/year/m² (passive system with polycarbonate), which was efficient at that time (the project dates from 1992).

The constructions' low thermal mass does not cushion temperature variations, which leads to risks of overheating that would be avoided by heavier masonry. On the other hand, it is easier to cool houses down at night for better sleeping conditions. This type of construction is therefore more suited to occupants who work outside the home during the day. In addition, the wooden material is renewable, which preserves natural resources. Manufacturers have reduced amount of formaldehyde in the adhesives used to assemble pressed wood.

The industrial wood-frame construction reduced the duration of the worksite. The technical skills of the workmen avoided problems of implementing this new technology (design and production of wood and aluminium frames, assembly of components according to OKALUX specifications). Capillary structures are delivered encapsulated between two panes, which ensures good protection on the worksite and is closer to standard double glazing. Due to a slight condensation phenomenon, which poses no major problem (slight drop in solar radiation transmission), the water evacuation holes were enlarged. The brick walls probably dry out over time.

Industrial production on a larger scale would bring down costs, which are currently 50 euro per m^2 for the raw material and 220 euro m^2 for the material encapsulated between two glass panes (OKALUX). Taking energy performance into account in the social housing price ceiling would make complementary investment possible, with amortization over time and generally lower combined rent + operation charges.

The Trombe walls are a little more aesthetic since the capillary structure reduces the visual impact of the black surface. The transparent insulation is an interesting material, both as an exterior insulator and as light-diffusing glazing. New possibilities are also available to architects to compose translucent façades and light interior environments, while avoiding glare and the cold wall effect.

5.1.1.6 Life cycle assessment

A life cycle assessment was carried out on these houses in comparison with a reference house, representative of a new construction in France. The reference was defined during a workshop organized by the Plan Urbanisme Construction et Architecture (urban planning, building and architecture programme, French ministry for housing). It involved an individual house of 112 m^2 on two levels. The techniques used (concrete blocks, polystyrene insulation from the inside, etc.) are currently the most widely used techniques according to a statistical study by INSEE. The results are presented in a radar chart (cf. figure 5.11 below). Each axis corresponds to an environmental theme. The indicators obtained for the project studied are given in values relative to the reference, placed at level 1.

Reducing heating requirements and using wood in building constructions diminishes most impacts (especially the greenhouse effect), which shows the advantages of this concept from an environmental point of view.

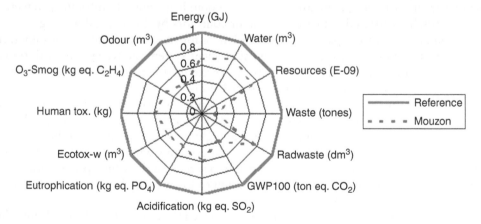

Figure 5.11 Results of the life cycle assessment.

However, technological innovations in windows (low-E layers, rare gases) have brought down the advantage of transparent insulation. In addition, research done in Germany has come up with the concept of a passive house that can reach higher performance levels at a similar cost (as we shall see later on).

In France, at the same time, decision-makers opted for a more general "high environmental quality" approach, which is interesting from a theoretical point of view but which has not tackled energy performance. The result was a fifteen-year time lag, and a lack of qualified people today to deal with the new energy crisis.

5.1.2 Experimental construction of Castanet Tolosan (Haute-Garonne)

This housing estate was constructed following a call for "high environmental quality" (HEQ) projects in late 1993 by the Plan Urbanisme, Construction et Architecture (PUCA, Ministry for Housing). The study presented here was carried out as part of a workshop to evaluate the environmental quality of buildings (ATEQUE), also organized by PUCA. It focuses on one of the houses in the estate.

The house was modelled in 4 thermal zones: each of the two floors is divided into two zones, one on the southeast side and other on the northwest. This makes it easier to see how orientation influences thermal performance (heating requirements and summer comfort). A fifth zone represents the garage, which is not heated. The simulation makes it possible to follow the temperature variation in this "buffer" zone and thus evaluate heat loss from heated zones. The heating requirements estimated by the COMFIE software are around 6000 kWh per year, i.e. around 70 kWh/m^2.

An 80-year time span was set for the analysis. Data collected thanks to the European project REGENER (Karlsruhe University and the Swiss base Oekoinventare) were considered for the following materials: bricks, mineral wool, exterior mineral coating, plaster board and tiles, concrete, tiling, flexible flooring, roof tiles, wood and paint. The manufacture and replacement of the windows (double glazing, PVC frame) were also taken into account. The lifespan of the windows and doors was set at 30 years and the wall coatings' lifespan at 10 years.

We compared the results of this analysis with those obtained for the standard house, defined by a working group at the same ATEQUE workshop (cf. preceding paragraph). The performances of both houses are very similar, cf. the chart below. The use of water-based paint reduced the impacts in terms of toxicity and smog, but for example the overall potential for climate change only differed by 2%.

This low performance gain is linked to the mediocre thermal quality. Many "HEQ" actors at the time did not assimilate the notions of bioclimatic design, preferring to focus on newer aspects (e.g. choice of materials). The quantified balances show the limits of this type of approach.

However, these performances could have been greatly improved. The orientation of the houses could have been chosen so that the widest façades (including the living room and kitchen) faced south with a high number of (low-emissivity) windows. The insulation could have been placed on the outside. Double-flow ventilation could have been considered. We now know that heating loads could have been divided by 5 (cf. examples given below), which would have greatly improved the environmental balance of these houses.

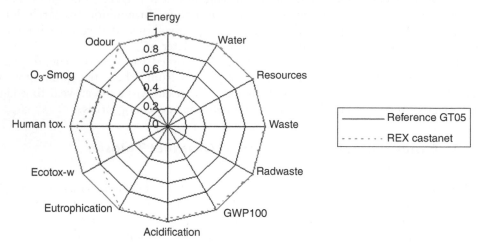

Figure 5.12 Comparative results between the Castanet Tolosan project and the regulatory reference at the time.

It is worth noting that the "bioclimatic" alternative requires a specific block plan study. An intern at the Toulouse School of Architecture verified whether this kind of redefined block plan would be possible in the case studied, which would also bring down construction costs if the housing was more compact (cf. the example of the solar town in Freiburg, Germany, described in more detail below).

5.1.3 Winner of the competition solar habitat, contemporary habitat (Rhone)

The winner of this biennial competition was a detached house, partially heated by direct solar flooring: solar collectors placed on the roof heat water, which then circulates under the floor. This one-storey house (cf. figure below) was modelled in four zones: the day zone (living room, kitchen), the night zone (bedrooms, workshop), the sunspace and the garage. A fifth imaginary zone was added to take into account materials not included in the thermal model: foundations, wooden beams and solar collectors. Since the house is located in the Rhône area (Monsols), a climate for the Mâcon region was considered at an altitude of 500 metres.

The house is moderately insulated (12 cm in the walls and roof), not very compact (only one level), and some of its windows face northwest (curved wall in the living room) or west (sunspace). The thermal performances are therefore fairly average: the heating load is 29,882 kWh, i.e. 132 kWh/m²/year. However, since 30% of the load is satisfied by the direct solar flooring (cf. diagram below), we have only considered 92 kWh/m²/year in the life cycle assessment. The choice of a one-storey house is penalizing when it comes to energy and the associated impacts, but the house has the advantage of being accessible to people with reduced mobility. This contradiction illustrates the variety of situations, and therefore the technical or architectural techniques chosen depending on the specific priorities of each project.

Figure 5.13 Winning house of the competition Solar habitat, contemporary habitat (1998), photo Systèmes Solaires[454], Architect: Atelier de l'Entre.

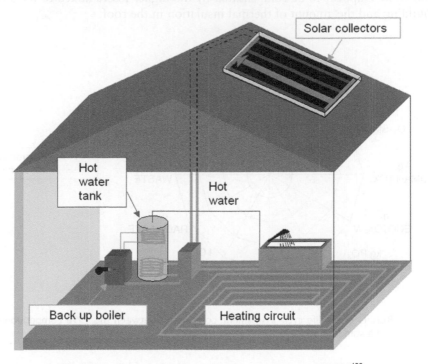

Figure 5.14 Diagram of direct solar flooring (source: ASDER)[455].

---

[454]Revue Systèmes solaires, www. energies-renouvelables.org
[455]Association Savoyarde pour le Développement des Energies Renouvelables, http://www. asder.asso.fr/

With an energy cost estimated at the time as 4 cents of a euro per kWh – lower heating value – (energy costs are taken from the journal Energie Plus) and average annual boiler efficiency of 80%, the cost of heating is around 4.5 euro/m², which is close to the performance indicated by the designers (4.1 euro/m²). By comparison, the requirements of the reference house (defined in ATEQUE, cf. above) are 70 kWh/m²/year, or 3.5 euro/m².

For the transportation of materials from their production site to the worksite, we considered an average of 20 km due to the local production of the dominant weight materials (i.e. stone).

The benefits of using wood gave us a negative GWP value for the building phase: −32 tonnes of CO_2. However, the emissions linked to heating are significant: 480 tonnes, which gives a fairly unfavourable overall balance for the duration of the analysis: around 2 tonnes of CO_2 emitted per habitable m², i.e. 25 kg/m²/year.

The results are presented as a radar chart (cf. figure below). For example, according to our estimation, the greenhouse effect contribution over the winning project's life cycle would be about the same as for a standard house. The benefits of the solar heating system appear to be compensated by the higher losses linked to the shape of the building and the amount of thermal insulation in the roof.

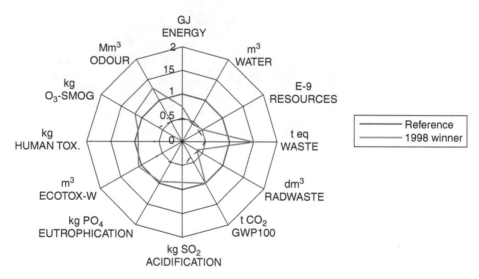

Figure 5.15 Results of the life cycle assessment, comparison between the 1998 competition winner and a standard house.

The CO_2 balance of wood is complex because of the variety of species, local forest conditions[456] and long-term carbon storage, which is not accounted for in the

---

[456] G.H. Kohlmaier, M. Weber and R.A. Houghton, *Carbon dioxide mitigation in forestry and wood industry*, Springer, 1998

standard LCA method[457]. Photosynthesis involves a great number of chemical reactions, but the production of cellulose can be represented by the following overall chemical equation[458]:

$$6\,CO_2 + 12\,H_2O + photons -> C_6H_{12}O_6 + 6\,O_2 + 6\,H_2O$$

To produce one kg of cellulose therefore corresponds to absorbing around 1.5 kg of CO_2. In fact, wood is composed of about 40% to 50% cellulose, but also 20% to 30% hemicellulose and 20% to 30% lignin. The carbon content is therefore around 0.5 kg per kg of dry wood, or 1.85 kg of CO_2 equivalent. Wood must be cut and transported, sometimes dried using combustibles, and sawn: these operations (directly or indirectly) emit CO_2, but in general (except for wood-based products with a high content of auxiliary materials, like plywood), the overall carbon balance gives a negative CO_2 content per kg of wood. On the other hand, the carbon returns to the atmosphere at end of life, as CO_2 and VOC at incineration, also producing methane (which may be recuperated) in the case of landfill. Wood can also be recycled or re-used. The end-of-life scenario (incineration with or without energy recovery, landfill with or without methane recovery, recycling, etc.) can thus influence the environmental balance over a building's life cycle.

Other authors consider an overall neutral CO_2 balance for wood, because they suppose that the CO_2 stored during photosynthesis ends up returning into the atmosphere. Work to compare these different approaches continues, in particular in European projects.

Since this study, new knowledge about forests has been developed, in particular by Lund and Uppsala Universities in Sweden. The carbon flows exchanged between trees and the atmosphere, and between the soil and the atmosphere, are measured using considerable resources (trees are encapsulated with plastic sheets). According to these measurements, it appears that the CO_2 flow absorbed by photosynthesis is similar to the flow emitted by the respiration processes of trees and soil. However, the forest considered in this study comprises adult trees (about one hundred years old) and low rainfall. On other sites, significant carbon immobilization was observed. According to the report by senator M. Deneux[459], the overall carbon balance still constitutes carbon sequestration in temperate forests.

The use of wood means that carbon can be stored during a building's life span, but this effect is not currently taken into account in balance sheets. Opting for certified wood, taken from a sustainably managed forest, should improve the environmental balance compared to a product that contributes to deforestation, since carbon is stored once again when a new tree grows. Methodological developments are therefore still required to make these balance sheets more precise.

---

[457] Ari Rabl et al., How to account for CO_2 emissions from biomass in an LCA, International Journal of LCA Vol. 12, No. 5, p. 281, 2007

[458] Frischknecht R., et al., 1996, Oekoinventare von Energiesystemen, 3. Auflage, ETH Zürich/PSI Villigen

[459] Marcel Deneux, *L'évaluation de l'ampleur des changements climatiques, de leurs causes et de leur impact prévisible sur la géographie de la France à l'horizon 2025, 2050 et 2100*, Office parlementaire des choix scientifiques et technologiques, January 2002

The two preceding examples show that despite the designers' proactive approach, the performances achieved are not very different from those of a standard house. It is therefore important to insist on the importance of combining reduced loads and energy production, preferably renewable, along with aspects linked to materials and energy. The life cycle assessment is an ideal tool for producing an overview of this type, whereas the 14 "targets" proposed by "HEQ" limit the choice possibilities between the phases of manufacturing the materials and using the buildings.

5.1.4 Passive houses in Formerie (Oise)

Almost 15 years after Germany's experiment, two passive houses were built in France in 2007 at Formerie, in the Oise area (North of Paris) by the company Les Airelles Construction. The project comprises two semi-detached houses with identical surface areas, orientations and lay-outs (except for the garage) on two floors (cf. image below).

Figure 5.16 Passive houses in Formerie (Oise), company Les Airelles Construction.

The Maison Passive label corresponds to very high energy performances (cf. chapter 4)[460], however, the French climate often brings the risk of overheating in summer. Quantifying the environmental impacts helps to evaluate the benefits of this approach, and the tools described in chapter 3 were therefore used to this end. The building's geometry was modelled using ALCYONE 2.1 software[461], 10 thermal zones were defined with, for each of the two houses:

– a living room zone + kitchen (day-time occupancy, south facing),
– an entrance zone + office (occasional occupancy, north facing),

---

[460] Annual heating load under $15 \, kWh/m^2$, air infiltrations less than 0.6 air changes per hour with a pressure difference of 50 Pa (measured by blower door test), cf. www.lamaisonpassive.fr and www.passiv.de

[461] Salomon T., Mikolasek R. and Peuportier B., Outil de simulation thermique du bâtiment, COMFIE, Journée SFT-IBPSA "Outils de simulation thermo-aéraulique du bâtiment", La Rochelle, 2005

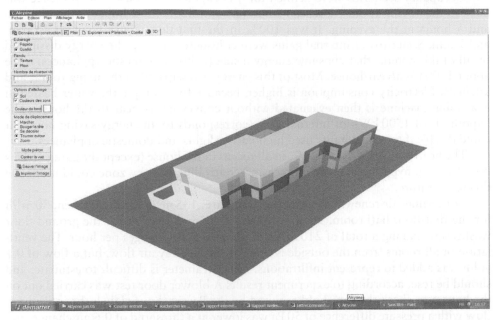

Figure 5.17 3D view of different thermal zones (ALCYONE software).

– a bedroom zone + lounge (night-time occupancy, south facing),
– a bedroom zone + bathroom (night-time occupancy, north facing),
– a garage zone, unheated.

The attic located above the insulation is considered to be at external temperature, and therefore this area is not modelled as a zone. The heated surface area of each house is 132 m².

Concern is sometimes expressed as to whether the passive house concept can provide thermal comfort during the summer, and even mid-season. South-facing first-floor zones are likely to benefit more from winter sun, but also to overheat in the summer. It is therefore worth paying particular attention to these zones when analyzing simulation results. Masks were specified on the southeast and northeast windows to take account of the presence of a barn and the unusual geometry of the southeast façade (presence of balconies, recesses). Efficient outside protection (external slat blinds) was installed, along with a ground-coupled heat exchanger. The solar protections were modelled with an 80% reduction of the windows' solar factor. The meteorological data file used to assess the risk of overheating in summer corresponds to the 2003 heat wave recorded in Montreuil (Paris suburbs).

Small thermal bridges were considered on all of the façades at the foundations and ceiling. The corresponding thermal loss coefficients per unit length were set at a fixed rate of 0.1 W/(m.K). It is difficult to evaluate these terms precisely because the heat transfer is highly dependent on the quality of construction work. This type of parameter needs to be calibrated according to experiment results. A alternative with doubled thermal bridges was evaluated for comparison.

Occupancy scenarios were defined for 4 occupants per house. The presence was 80% in the "living room + kitchen" zones (ground floor) for 2 hours in the morning and 5 hours in the evening. It was 100% in the first-floor zones between 10 p.m. and 7 a.m. Scenarios of internal gains were considered, taking the energy dissipated by all of the systems that consume energy inside (lighting, domestic appliances) to be about 1500 kWh/year/house. Most of this energy is dissipated in the living rooms and kitchens. Electricity consumption is higher, because for example the water heated in a washing machine is then evacuated without conveying its heat to the house. The hypothesis of 1500 kWh of internal gains corresponds to an energy-saving house. It could double if the inhabitants used high-energy lights and domestic appliances.

The heating is set at 19°C, 24/7, in all rooms in the house (except the garage). This is of course a hypothesis: using a timer and controlling heat by zone could modulate this temperature.

Concerning air renewal, healthy flow rates are: 135 m^3/h for the kitchen, 30 m^3/h for the first-floor bathroom, 30 m^3/h for the toilet and 15 m^3/h for the ground-floor washroom, making a total of 210 m^3/h, or about 0.5 air changes per hour. The ventilation of all rooms from the outside is fixed at this healthy air flow, but a flow of 0.1 vol/h was added to represent infiltrations. This parameter is difficult to estimate, and should be reset according to experiment results A blower door test was carried out on the houses to check the threshold required by the Passive house label: the infiltration flow with a pressure difference of 50 Pa was lower at a threshold of 0.6 air changes per hour. However, this pressure difference varies over time (depending on wind and temperature), and therefore uncertainty remains as to the average flow during a heating season, taken into account in these calculations.

The balanced ventilation exchanger was taken to be 80% efficient for heat recovery, but because the air entering by infiltration is not pre-heated, an overall efficiency of 70% was taken into account over the 0.6 vol/h. Over-ventilation of 10 vol/h during the night, corresponding to open windows, was considered for all rooms in the house (except the garage) from 11 p.m. to 8 a.m. in the summer. The balanced ventilation exchanger is bypassed so that it does not pre-heat fresh air in the summer.

The annual heating load estimated by these calculations is lower than the 15 kWh/m^2 threshold required by the label, but significant uncertainty remains regarding thermal bridges, air infiltrations and the occupancy scenario. The heating period runs from the second week in November to mid-March. A simulation with thermal bridges that are twice as big results in about 20% more heating load than the preceding simulations.

An alternative design was tested to evaluate the heating load's sensitivity to the window area on the southeast façade. In this alternative:

- The French windows measuring 1.80 m × 2.15 m in bedroom 2 (1st floor) were replaced by French windows measuring 0.90 m × 2.15 m, i.e. half the area;
- The French windows furthest to the east in the living room (ground floor), measuring 180 m × 2.15 m were replaced by windows measuring 0.9 m × 1.20 m, or a division by 3.5 of the area.

The heating load calculated thus increases by around 11.5%: even when a highly insulated envelope brings down the heating load, solar gains are useful, and wide, south-facing bay windows are generally more efficient than small windows.

For the comfort study, simulations were carried out for a typical year (in Trappes), as well as for a heat wave to determine whether the houses remained comfortable during extreme weather. During the typical summer, and provided inhabitants behave suitably (i.e. close solar protections, open windows at night), the temperatures in the houses remained moderate: even in the most exposed zone (west house, first floor, south), the number of hours during which the temperature exceeded 21°C was around 100: increasing the thermal mass of the houses does not appear to be necessary.

During a heat wave lasting about ten days, the building appears to resist well, since the temperature of the most exposed zone (i.e. west house, first floor, south) remained around 10°C below the outside temperature (with external rolling blinds and a ground-coupled heat exchanger). The ground-coupled heat exchanger was evaluated using a model with finite volumes[462]. According to these calculations, it reduces the maximum temperature of the houses by 2.5°C (cf. the graph below).

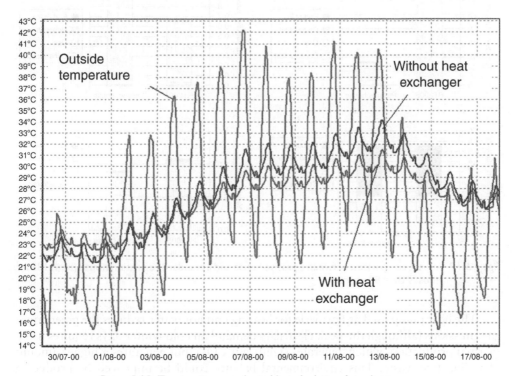

Figure 5.18 Temperatures evaluated by calculation for a heat wave.

In noisy zones, it is not possible to open windows at night. In such cases, a ground-coupled heat exchanger is even more worthwhile. According to these calculations, for equivalent solar protection, a passive house is at no greater risk of overheating during a heat wave than a house built to current regulation standards. The reinforced insulation

[462]Stéphane Thiers and Bruno Peuportier, *Modélisation thermique d'un échangeur air-sol pour le rafraîchissement de bâtiments*, Themed Day SFT-IBPSA-France, Aix-les-Bains, April 2007

protects as effectively from the heat as from the cold, it slows down exchanges and means that night-time heat can be retained by over-ventilation (night temperatures ranged from 21°C to 25°C in 2003, cf. above graph).

A passive house employs more materials than a standard house (thicker insulation, triple glazing, etc.). A life cycle assessment can evaluate the overall benefits of this concept, taking into account the additional impact of manufacturing products and the energy saved. The graphs below show a comparison with an identically shaped building using technologies corresponding to regulation reference values and gas heating[463]. The passive design reduces most of the environmental impacts: the increase in impacts linked to manufacturing the materials is largely compensated by the impacts avoided due to energy savings, considering an 80-year life span in this analysis.

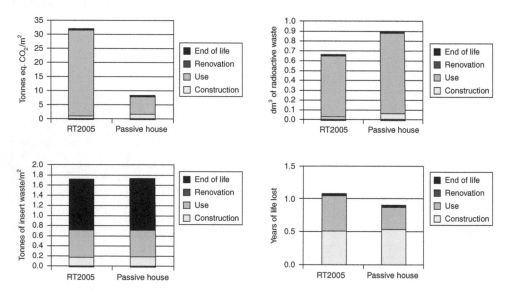

Figure 5.19 Impact comparison between a passive house and a standard regulation house (both 132 m² houses over 80 years).

The heat pump used at Formerie for heating and hot water reduces the greenhouse gas emissions (although the peak electricity demand on cold winter days requires using thermal production, which generates significant CO_2), but generates a higher amount of radioactive waste. This environmental balance could be improved by producing renewable energy, e.g. with a photovoltaic roof like on the ZEN house[464] or the Plus Energie houses (cf. §5.3).

In this balance sheet, which includes all the energy consumed in the building (including lighting and domestic appliances), the construction phase represents a third of greenhouse gas emissions. The materials constitute the majority of the waste

---

[463]Stéphane Thiers and Bruno Peuportier, *Thermal and environmental assessment of a passive building equipped with an earth-to-air heat exchanger in France*, Solar Energy, Vol. 82 (9), pp. 820–831

[464]Alain Ricaud et al., cf. http://www.cythelia.fr/

produced (especially in the demolition phase), and contribute significantly to the toxicity (years of life lost – DALY, especially during the manufacturing phase).

The ratios expressed per m^2 and per year can be used to compare different projects, cf. table below.

Table 5.6 Detached houses, ratios per m^2 and per year.

Indicator	Passive house heat pump	Ref. regulation gas heating
Primary energy (MJ/m^2/yr)	330	633
GWP (kg CO$_2$/m^2/yr)	7	27
Radioactive waste (cm^3/m^2/yr)	0.75	0.55
Other waste (kg/m^2/yr)	18	18

This balance could be improved in several ways, e.g. by producing energy locally from renewable sources, recycling materials at end of life, saving water and/or collecting rainwater. However, a more general question, raised above, concerns whether individual houses are a viable option compared to collective housing that consumes less materials and energy due to compactness, and reduces transport requirements and urban sprawl. In addition, the examples above correspond to new buildings, which only represent 1% per year of the existing building stock. The following paragraphs give examples of apartment buildings, retrofit projects, and eco-neighbourhood experiments.

5.2 APARTMENT, TERTIARY BUILDINGS AND RETROFIT

5.2.1 Tertiary building in Mèze (Herault)

This example concerns a project envisaged in Mèze (Hérault). The clients' objective was to reduce the emissions over the building's life cycle by a factor of 4. This factor of 4 is based on the following hypotheses: emissions generally need to be halved (on a world scale), in particular to combat the greenhouse effect, yet the world's population is set to grow by a factor of two, making it necessary to reduce by 4 in comparison with current standards.

The building "Le Nautile" is organized in a semi-circle running from the southeast (greenhouse) to the southwest, to make the most of solar gains (cf. figure below). The windows are protected from the summer sun by large overhangs that act as a light shelf.

The building was compared to a typical tertiary building in the region of Montpellier, with the same useful area, i.e. 1870 m^2. The reference wall compositions are made of concrete (20 cm thick) insulated from the inside (8 cm of extruded polystyrene), as opposed to a wooden frame insulated with 10 cm of recycled cellulose in the project studied. The reference roof terrace is insulated by 9 cm of polystyrene, compared to 16 cm of cellulose in the project. The reference building is equipped with standard double glazing, while the project includes low-E double glazing and its thermal mass is reinforced by adobe walls between the greenhouse and the offices. The air exchanges between the greenhouse and the other zones were taken into account.

The heating and cooling loads are given in the table below (COMFIE software results). The evaluation was compared to the results of the TRNSYS software applied

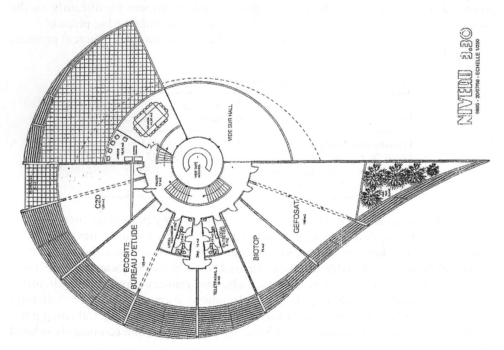

Figure 5.20 Tertiary building project, Architect: Gilles CHICAUD.

Table 5.7 Heating and cooling loads of the two tertiary buildings.

	Annual heating load (kWh/m²/yr)	Annual cooling load (kWh/m²/yr)
Reference	36	50
Le Nautile	15	6

by EDF on the same building. The difference was 2% for the heating load and 9% for the cooling load (the homogeneity of the data considered in the two studies was not verified for all of the parameters).

The sources of impact taken into account in the LCA using the EQUER software are heating, lighting and ventilation, along with building materials (manufacture and building, renovation and demolition). The comparative ecoprofile obtained is given in figure 5.16 in relative values.

To compare the results with those obtained with other tertiary buildings, ratios per m² and per year are given (table below) for several impact indicators over the whole life cycle.

It is interesting to observe that good thermal design leads to significantly less environmental impacts (about factor 3 in this project, not 4), bringing additional arguments to convince local decision makers of the merits of this approach.

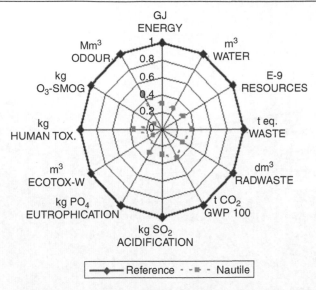

Figure 5.21 Tertiary buildings, comparative ecoprofile.

Table 5.8 Tertiary buildings, ratios per m² and per year.

Indicator	Le Nautile	Reference
Primary energy (MJ/m²/yr)	520	1720
Building materials (kg/m²/yr)	5.7	15.5
GWP (kg CO₂/m²/yr)	6.8	18.4
Acidification (kg SO₂/m²/yr)	0.03	0.1
Smog (kg C₂H₄/m²/yr)	0.006	0.06

5.2.2   Retrofit of a social housing apartment block in Montreuil (Paris suburbs)

The project is located in Montreuil (Paris suburbs) in the La Noue neighbourhood comprising collective apartment blocks (4 to 18 stories), in which the municipality has set up an "environment neighbourhood" pilot project. The building chosen for the experiment was a small 4-storey block built in 1969 with an exposure ideal for taking advantage of solar gains. The operation was carried out by the local social housing office (OPHLM de Montreuil) with the support of the European Commission and ADEME[465]. The client's main aim was to improve the image of the neighbourhood and reduce the operating charges.

This operation, designed in 1998, was carried out from 2001–2002 as part of the European project REGEN LINK. The project was coordinated by the Dutch organization PATRIMONIUM, which brought together eight social housing

---

[465] French Environment and Energy Management Agency, www.ademe.fr

Figure 5.22 General view of the building after retrofit, south façade.

organizations to produce innovative demonstration projects in the domain of energy efficiency and renewable energy. The operations were part of urban projects aiming to improve neighbourhoods' environmental quality.

The retrofit integrated bioclimatic and life cycle aspects, in particular: external insulation of façades using 10 cm glass wool (instead of 6 cm for standard renovations), and a siding made from recycled materials (cellulose, plastic). The existing single-glazed windows were replaced by double-glazing including a low-emissivity coating that reduces heat loss by radiation. Glazed balconies, argon-filled glazing, which provides even better insulation, hygro-adjustable ventilation (which reduces the air flow if the apartment is unoccupied), and water-saving showers and taps were installed in some apartments. A district heating system provides heating and hot water for the building.

The south-facing living rooms (bedrooms are north-facing) meant that "bioclimatic" concepts could be implemented: the glazed surface is larger on the southern façade, and fresh air is pre-heated in glazed balconies. These glazed balconies also allow thermal bridges to be reduced (the balcony floor slabs interrupt the insulation). Since the apartments look out in two directions, they are easy to ventilate in summer time.

Calculations by dynamic thermal simulation were carried out to choose the most suitable type of glazing, size the insulation, and study summer comfort. The low-emissivity coating on the windows is more transparent on the south side ("hard" pyrolitic coating), to transmit the solar radiation better, and more insulating on the other façades ("soft" coating magnetron sputtering vacuum deposition). According to these calculations, heating requirements are about $85\,kWh/m^2/year$ for a temperature of 20°C in the apartments. In fact, the inside temperature is often higher than 22°C, which explains the level of $100\,kWh/m^2/year$ measured. The control of the heating system could therefore be slightly modified to improve performance further still.

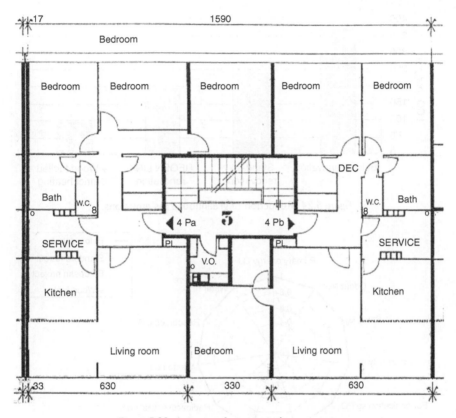

Figure 5.23 Lay-out of two typical apartments.

The local Montreuil Vincennes Energy agency took part in the work by informing inhabitants. Individual diagnoses of electricity demand management were also carried out and were a way of making tenants aware of energy savings that could be made using low-energy lamps and domestic appliances. A measurement campaign evaluated the consumption of the different items (lighting, refrigeration, television, washing, etc.). The electricity saving potential was evaluated at around 40% of the bill, i.e. around 1150 kWh/year/apartment, spread over three main items: refrigeration, lighting and standby. This confirms earlier research done by Olivier Sidler in the Drôme area[466].

The reduction in environmental impacts was assessed by a life cycle assessment. The graph below shows the reduction in greenhouse gases for a standard renovation (6 cm insulation on the façade and replacement of windows with standard double-glazing), the project presented here, and a alternative in which the district heating system is fuelled by wood.

The CO_2 emissions would be much lower if the district heating system were fuelled by wood instead of the current fuel and coal mix.

---

[466]Olivier Sidler, Connaissance et maîtrise des consommations des usages de l'électricité dans le secteur résidentiel, May 2002, cf. www.enertech.fr

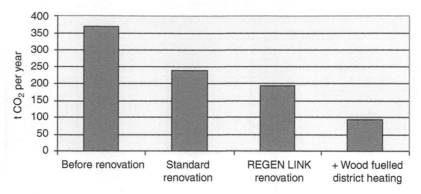

Figure 5.24 Reduction of greenhouse gas emissions.

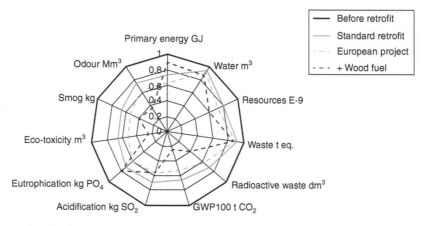

Figure 5.25 Comparison of environmental impacts before and after retrofit.

A sensitivity study focused on the insulation thickness on the façades. The first centimetres are very efficient in reducing heating loads, but the benefits of an additional centimetre then diminish (cf. figure below). On the other hand, the impacts linked to manufacturing the material are proportional to its thickness. The accumulation over the life cycle therefore includes an optimum, but it is very flat: there is very little variation between 20 cm and 40 cm.

A one-year measurement campaign was performed to assess the efficiency of these solutions: despite an increase in temperature observed in the apartments, the heating savings were over 30%. The thermal and acoustic comfort was improved thanks to the insulation of the façades and double-glazing. The following graph shows the progression of temperatures during the 2003 heat wave.

The high thermal mass generated by the floor slabs and concrete walls attenuates temperature variations: at the hottest point of the heat wave, the inside temperature remained between 8°C and 10°C below the outside temperature. The large amount of glazing (40% on the south face) takes advantage of daylighting. This operation avoids

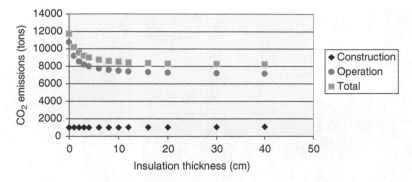

Figure 5.26 CO_2 emissions depending on insulation thickness on façade.

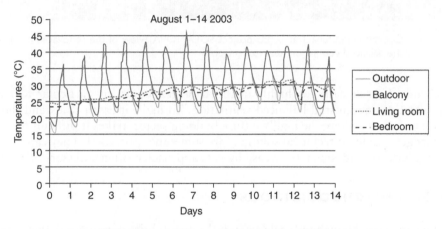

Figure 5.27 Temperatures measured during the 2003 heat wave.

emitting 76 tonnes of CO_2 each year. It cost 185,000 euro, or about 3500 euro per apartment. Overall, the turn-around time is 15 years, but some technologies are more economical than others (choice of more efficient glazing, water-saving devices).

More recent projects have taken performance levels higher, at times approaching the level of passive houses. For example, in Hungary, an operation was carried out in Dunaujvaros as part of the European project SOLANOVA[467]. By insulating the façades, replacing the windows and installing balanced ventilation with heat recovery, the annual heating load was quartered, dropping to below 40 kWh/m^2.

Even higher performance levels have been achieved in Germany, e.g. in Nuremberg, where heating requirements have been reduced by 87%, or 27 kWh/m^2 after retrofit[468].

---

[467] Andreas Hermelink, *Reality Check: The Example SOLANOVA, Hungary*, European Conference and Cooperation Exchange, Sustainable Energy Systems for Buildings – Challenges and Chances, Vienna, Austria, November 2006, cf. www.solanova.org
[468] Uli Neumann, EU project TREES, Section 3 case study, 3.5 Nurenberg, http://direns.mines-paristech.fr/Sites/TREES, 2007

Figure 5.28 Operation in Dunaujvaros (EU project SOLANOVA), view of the renovated building and the balanced ventilation heat exchanger[469].

This kind of result calls for highly qualified professionals, in particular to master air infiltrations and install balanced ventilation. To make it easier to extend thermal retrofit in France, where the highly diffuse building sector makes it difficult to spread knowledge, simplified technical solutions are sometimes proposed. They can supply a working base for tradesmen, but a case-by-case approach is often necessary on the field.

5.3 ECO-NEIGHBOURHOODS

Working on a neighbourhood scale extends the possibilities of reducing environmental impacts, but calls for more complex analysis. Knowledge is less refined, which leaves more room for subjectivity. The risk thus arises of reproducing the errors of early "HEQ", i.e. getting the priority wrong by putting more emphasis on visible elements (building materials, the plume produced by steam coming out the chimney, worksite dust, etc.) and ignoring the effects (toxicity, biodiversity), replacing technical skills by procedures with random efficiency (certification), or not measuring the performances obtained to learn from experience, etc.

We present here several flagship operations and sketch out a methodology for rationalizing approaches.

5.3.1 Écolonia project (Netherlands)

To build with an environmentally responsible, energy-saving approach requires knowing what is possible in practice. To advance knowledge, a demonstration project of 100 ecological apartments was constructed from 1991 to 1993 in Alphen aan den Rijn, a new town in the west Netherlands.

---

[469]Tamas Csoknyai, EU project TREES, Section 3 case study, 3.2 Dunaujvaros, http://direns. mines-paristech.fr/Sites/TREES, 2007

Following a selection, nine architects designed housing in line with the draft urban development plan. The guidelines followed three rules from the 1989 Netherlands National Environment Plan:

- Energy saving;
- Management over the life cycle, use of recycled or recyclable materials, or even re-usable materials, with low-impact manufacturing and end-of-life management;
- Improved construction quality.

ECOLONIA's aim was to be a demonstration project rather than an experimental field. This therefore meant tackling themes already defined, e.g. show that it is possible to significantly reduce energy consumption for heating and hot water. An additional objective was to reduce the amount of "grey" energy, i.e. contained in materials. This contribution to the energy balance increases as energy loads drop during the use stage.

5.3.1.1   Description of the programme

Following numerous projects on managing energy or specific environmental issues, an overall approach was sought. The idea was to take advantage of knowledge and general experience in the field, and supply new information to those architects, construction companies and consultants who need the knowledge. The project was set up very quickly and then extended.

The 1989 Netherlands National Environment Plan played a crucial role in these discussions. The plan centred on energy saving and protecting the environment including targets, in particular for the building sector. ECOLONIA's object was to put these recommendations into practice to encourage decision-makers to develop similar projects. The three measures recommended by this plan (energy efficiency, management over the life cycle, and improved building quality) were the working base for putting together the urban plans and designing the architecture. For the architectural aspect, each of the preceding measures was divided into three sub-sections.

Energy saving

This involves reducing the consumption of fossil energy by reducing loads, making use of renewable energy and installing energy-saving heating and ventilation equipment. The sub-sections are:

- Reducing heat loss;
- Using solar gains;
- Reducing consumption over the construction and use stages.

Management over the life cycle

This term refers to the life cycle from the extraction of raw materials up to the demolition and possible re-use. The main aspects concern the effects on the environment and the energy demand of the processes and products used. The sub-sections relating to architectural quality are the following:

- Limiting water consumption and recycling materials;
- Using sustainable materials, requiring little maintenance and/or of agricultural origin;
- Ensuring flexible construction and occupancy.

Improved construction quality

A product is said to be of "good quality" if it fulfils a given function over a long period, if it can be repaired and re-used, and if it does not generate waste that harms the environment. Occupants must not suffer from any adverse consequences as a result of polluting emissions during the use stage, an aspect which requires paying careful attention to the indoor environment. The sub-sections relating to architectural quality are the following:

- Improving acoustic comfort;
- Taking health and safety aspects into account;
- Constructing with a "bio-ecological" approach.

Each of the nine architects had to respect the basic programme, plus one of the nine sub-sections mentioned above. In addition, the client (the Netherlands municipalities fund for building work) imposed strict cost requirements on the building.

The basic programme comprised the following requirements:

- energy consumption under $300\,MJ/m^3$ (about $35\,kWh/habitable\,m^2$);
- solar water heaters;
- boilers emitting low levels of NO_x with over 90% efficiency;
- loss coefficient under $2\,W/(m^2.K)$ for the windows;
- environmental quality considerations when choosing materials for outside drainage and inside pipes;
- crushed concrete as an aggregate;
- no tropical hardwoods;
- anhydrite floors;
- water-savers;
- environmental quality considerations when choosing wood treatment products;
- no CFCs in materials and processes;
- household waste sorting;
- no bituminous products;
- no radon in the housing.

For the three sub-sections connected to saving energy, the maximum energy consumption was set at $220\,MJ/m^3$ instead of 300 (i.e. about $25\,kWh/habitable\,m^2$).

5.3.1.2   Energy-saving design

The apartments were designed to receive solar radiation, for both heating and day-lighting (light reaches through to the centre of the apartments). A solar water heater was integrated into the roof. The boiler and (balanced) ventilation system were placed in a core zone.

The children's bedrooms are partially in the basement, so as to reduce heat loss. The heating control system takes the three thermal zones into account. In some of the apartments, an active solar system provides some of the heating in addition to hot water.

The additional cost of improving the environmental quality came to 10% of the total cost of the construction.

5.3.1.3 Managing over the life cycle

An estimate of the energy flows over the life cycle of the buildings was carried out, based on a 75-year life span and two renovations during that period. The energy required to build a house was evaluated at between 200 and 400 GJ, and the renovation was taken to consume 200 GJ. Concerning the use stage, the different annual flows (expressed in primary energy) are as follows:

- central heating: 14 GJ
- domestic hot water: 14 GJ (without solar heater)
- cooking: 2 GJ
- electricity: 15–25 GJ

Over the entire life cycle, the total ranges from 3775 to 4725 GJ. A solar water heater reduces domestic hot water requirements by about 40%.

The table below, provided by the Netherlands Agency for Energy and the Environment (NOVEM), shows the impacts linked to window joinery for various materials.

Table 5.9 Comparison of window joinery impacts.

	Aluminium	European wood	Iroko wood	Meranti wood	PVC
gas (m³)	372	200	233	235	241
coal (kg)	153	45.6	48.8	50.3	35.6
fuel oil (kg)	67	21.6	17.3	22.6	30
total (MJ)	23 195	10 774	12 086	12 444	12 583
m³ of tropical wood	3.1	0.6	222	39.8	1.1
CO_2	794	443	525	524	514
waste special (kg)	27.6	17.8	18.1	16.2	16.1
total waste (kg)	520	203	415	221	162

Techniques exist to reduce energy needs still further, e.g. closing insulating shutters at night, or recovering energy from waste. It is also important to ensure satisfactory comfort in summer time. One of the project's conclusions was that the most commonly used boilers are unsuitable for very low heating loads.

An every-saving scenario was defined to correspond to the plan's objective of reducing household electricity demand in the Netherlands by 40%. This scenario anticipates replacing electric heating and air conditioning with gas, and reserving electricity for lighting, force and electronics. The consumption of household appliances is given in the table below. In the "energy-saving" scenario, electricity is partially replaced by gas (e.g. to heat washing-machine water).

The main conclusions of these evaluations are as follows. It is possible to quarter an apartment's primary energy consumption using technology available on the market, split as follows (primary energy):

- central heating   2000 kWh/year (instead of 9000)
- domestic hot water   1400 kWh/year (instead of 3000)
- electricity   1600 kWh/year (instead of 7000)
- TOTAL   5000 kWh/year (instead of 19,000)

Table 5.10 Energy consumption of household appliances.

Appliance	1985, standard (kWh/an)	1991, best technology, electricity (kWh/yr)	E-saving scenario, electricity (kWh/yr)	E-saving scenario, gas (m³/yr)
Clothes washer	280	13	95	14
Clothes dryer	135	135	11	16
Dish washer	42	39	10	6
Hair dryer, iron, coffee machine	109	108	0	0
Refrigerator/freezer	639	326	0	65
Solar water heater	–	–	40	−120(*)
Other	1643	832	449	0
Total	2848	1653	605	−19
Primary energy (kWh)	7000	4200	<1600	

(*) the solar water heater saves on gas

This approach could be taken further still to achieve energy self-sufficiency using individual cogeneration systems or renewable energy.

5.3.2 Vauban neighbourhood in Freiburg (Germany)

The Freiburg region enjoys a better climate than most other parts of Germany, added to which local policies promote environmental protection, with a particular focus on renewable energy. Numerous initiatives have been set up, associating research (the Fraunhofer Institute for solar energy systems, one of the most advanced in the world in this area), industry (solar equipment producers, like "Solarfabrik"), the municipality, and final users with the "solar city" project.

When the French army left Germany in 1992, three years after reunification, the Vauban neighbourhood that housed the barracks was restructured. The local authorities opted to take a global approach integrating technical aspects (passive house label, positive energy houses, rainwater management, priority for public transport, cyclists and pedestrians, etc.), economic aspects (life cycle cost approach, anticipation of environmental shocks rather than emergency management, etc.), and social aspects (social housing, participative decision-making, tenant cooperatives, etc.). The participative process brought together current inhabitants – including squatters of the old buildings – and future inhabitants. It lasted three years during which time pollution was removed from the former military terrains.

It would take too long to describe here all of the innovative achievements in the neighbourhood and the city, which are well worth a visit: a specialist organization gives technical guided tours[470]. In the new building domain, the "solar estate" of 59 Plus Energy houses designed by the architect Rolf Disch, is without doubt one of the most efficient today. The heating loads are limited thanks to the "passive house" concept (cf. §4.1.1), and the Fraunhofer Institute have measured annual values of

---

[470]Cf. www.freiburg-futour.de, www.freiburg.de, www.solarregion.freiburg.de

Figure 5.29 Plus Energy houses in Freiburg (architect: Rolf Disch).

11 kWh/m^2 for this type of housing[471]. Photovoltaic roofs produce more energy than that required by the houses, and the same goes for the neighbouring social housing. The architect gives the following balance sheet for a 137 m^2 two-storey house (for a family of 4)[472]: consumption for heating and hot water 3550 kWh, and 2300 kWh for electricity, photovoltaic production 1000 kWh per kWp (peak kW, cf. chapter 4). The architecture, which optimizes compactness and orientation, makes it possible to attain performances at a market-compatible cost (2000 euro per m^2): all of the houses were sold, and the estate was later extended.

On a more collective scale, the energy consumption of a building containing car parks and a supermarket is ensured by a photovoltaic system integrated into the roof. The former barracks were transformed into student residences and flats, and a community centre houses cultural and administrative activities. Because the low-energy approach and passive house label apply to the whole neighbourhood, heating loads are minimal, with domestic hot water partly supplied by thermal solar systems.

---

[471] Volker Wittwer and Karsten Voss, *Solar-Passivhaus "Wohnen & arbeiten"*, *Freiburg, Vauban*, Förderung durch die Deutsche Bundesstifftung Umwelt, Förderkennzeichen 12333, Freiburg, March 2001
[472] La Maison écologique No. 46, August–Sept. 2008, www.lamaisonecologique.com

Additional heat is provided by a district heating system whose boiler is fuelled 80% by granulated wood and 20% by natural gas. Photovoltaic solar power supplies 65% of the neighbourhood's electricity requirements.

Retrofit operations have also been carried out; the example below shows how photovoltaic components have been integrated.

Figure 5.30 Photovoltaic systems integrated into an existing building in Freiburg (Germany).

In addition to energy aspects and the environmental quality of the buildings (e.g. using rainwater for the garden and toilets, filtering wastewater through plants), the urban space is designed to include numerous green areas (biotope reserve along a river, urban garden allotments, etc.), with priority for pedestrians and cyclists. Outside the main tram axis, speeds are limited to 30 km/h, and even 5 km/h in walkways leading to the buildings. Children can therefore safely go to school on bicycles. All of the services (shops, schools, tram) are located no further than 700 m from the apartment blocks, which reduces the need for vehicles (only 9% of households have a car).

The neighbourhood, which covers 37 hectares, groups about 5000 inhabitants, who are generally fairly wealthy despite the inclusion of social housing. A second similar neighbourhood, Rieselfeld, was built later with more social housing and an old people's home.

5.3.3 Lyon confluence neighbourhood

The Lyon Confluence project concerns the redevelopment of the whole peninsula that runs between the Rhone and Saone Rivers beyond Perrache Station, covering almost 150 ha. The study area selected represents around 60,000 m² of housing and 15,000 m² of offices. It also includes numerous public areas (about 70,000 m² of green spaces, streets and quays).

The project included a life cycle assessment, in which the method described in chapter 3 was extended to a neighbourhood scale[473], incorporating:

- Different types of building (houses, flats, shops, offices, schools, hospitals, hotels, etc.),
- Access infrastructure (roads and streets, car parks, green areas, etc.),
- Networks (water supply system, sewers, waste management, district heating, etc.).

The model also takes into account aspects linked to residents' behaviour (water and energy consumption, waste treatment, percentage of sorting and recycling, etc.) and the site's characteristics (transport distances, climate, type of energy source used to produce electricity and district heating, etc.). The tool developed can be used in different situations.

Construction of a new neighbourhood on an undeveloped site. The model shows how design choices influence the foreseeable impacts during the given analysis period (e.g. 80 years). It includes the manufacture, maintenance and replacement of the components (e.g. the choice of more durable components can reduce "refurbishment" impacts), and their end of life (landfill, incineration, etc.), and also the operation stage (choosing a more compact urban form can reduce heating loads and the corresponding impacts). Design alternatives can be compared to guide decisions.

Demolition of an existing urban block and construction of new buildings. The calculation can be carried out in two phases: first modelling the old urban block (only the end-of-life stage is taken into account), then the new project (cf. the case above).

Study to compare a retrofit of an existing urban block with its "renovation" (demolition then reconstruction). The demolition + reconstruction calculation corresponds to the above case. Retrofit also requires a double calculation. A useful method is to make an inventory of existing buildings, then carry out a second calculation to evaluate the projects after retrofit (it is once again interesting to compare several alternatives). The impacts of the actual retrofit can be deduced by subtraction, and the second calculation can evaluate subsequent impacts (i.e. operation, refurbishment and end of life).

---

[473]Emil Popovici, Contribution à l'analyse de cycle de vie des quartiers, doctoral thesis, Ecole des Mines de Paris, 209p, 2006

A combination of these different situations can be envisaged in a single neighbourhood. This involves splitting the neighbourhood into sub-sections and carrying out the preceding analyses on each of them.

The indicators described in chapter 2 are evaluated for each type of building in the neighbourhood over its life cycle, then over the neighbourhood as a whole. Creating a radar chart with 12 axes associated with these 12 indicators, diverse alternatives can be compared to improve the project from an environmental point of view. The new ARIADNE software supplements a set of tools to be used for buildings:

- ALCYONE, geometric description of buildings,
- COMFIE and PLEIADES[474], thermal simulation (cf. §3.2)
- EQUER, LCA of buildings (cf. §3.1).

These tools are chained via text files in which the building data (geometry, energy needs, quantities and types of materials used) are written by a tool located upstream and read by a tool located downstream. The new ARIADNE software is linked to EQUER, thus prolonging the data transfer chain. The EQUER results are imported for each type of building, stipulating their number. Additional data need to be supplied on the outside areas and networks, so that an LCA of the neighbourhood can be carried out.

Using the model described above, the performances of the Lyon Confluence project were compared with two references: one corresponding to the current standard and the other to best practices[475]. Three alternatives were thus defined:

- The Standard alternative roughly corresponds to regulations for new buildings (RT2005) and techniques generally used in France;
- The Base alternative corresponds to the technical and architectural choices made by SEM Lyon Confluence to meet the objectives of the EU Concerto programme (−40% energy consumption compared to RT2000, covered by renewable energy for 80% of heating and hot water, and 50% of electricity consumption of common areas);
- The Best Practices alternative corresponds to the most efficient existing techniques (generalized use of balanced ventilation with heat recovery, high insulation, thermal bridge treatment).

Thermal study

The twenty buildings in the neighbourhood studied (blocks A, B & C) were modelled considering several thermal zones, and taking into account masks generated by the neighbouring buildings see figures 5.31, 5.32 and 5.33 below.

The hypotheses common to all of the alternatives are: a constant 20°C thermostat set point; internal gains of $21\,\text{kWh/m}^2$ ($25\,\text{kWh/m}^2$ from which we remove washing-machine and dishwasher consumptions, since the hot water is evacuated); and the

---

[474]Salomon T., Mikolasek R. and Peuportier B., Outil de simulation thermique du bâtiment, COMFIE, SFT-IBPSA day, "Outils de simulation thermo-aéraulique du bâtiment", La Rochelle, 2005, cf. www.izuba.fr

[475]Bruno Peuportier, Emil Popovici and Maxime Trocmé, Analyse de cycle de vie à l'échelle du quartier, ADEQUA Seminar, Chambéry, October 2006

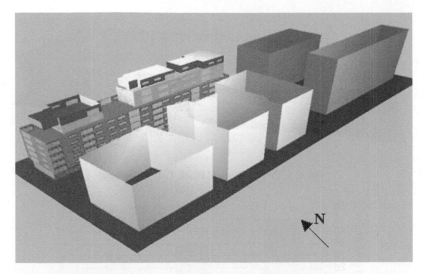

Figure 5.31 Block A, modelling of the north part (the colours correspond to the different thermal zones of building D).

Figure 5.32 Block A, modelling of the south part.

occupancy of the residents: 25% occupancy from 8 a.m. to 6 p.m. during the week and 100% the rest of the time (each resident emits 90 W of heating power).

The three alternatives differ in their techniques (insulation, ventilation, etc.), but the architecture also influences performance: for identical technical characteristics, the consumption calculated by simulation varies from single to triple between buildings 1 and 10 in block B for the Best Practices alternative (ratio of 4 kWh/m² for building 10 compared to 13 kWh/m² for building 1), and almost single to double (29 kWh/m² and 58 kWh/m²) in the Standard alternative.

The technical specifications, which fulfil the requirements of the EU Concerto programme, limit heating consumption to 60 kWh/m² for housing and 40 kWh/m² for offices, making an average 56 kWh/m². The results obtained using the COMFIE

Figure 5.33 Modelling of block C, housing side.

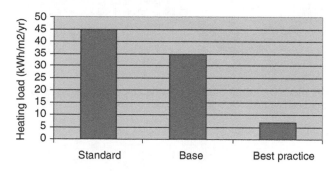

Figure 5.34 Heating loads for the 3 alternatives.

software correspond to this objective (on average over the 3 blocks), taking an average 60% efficiency for the wood-fired boiler.

Life cycle assessment

The buildings are heated with 100% gas in the Standard alternative, and 80% wood and 20% gas in the Base and Best Practices alternatives. The annual electricity consumption (78% nuclear production, 14% hydroelectricity, 4% gas and 4% coal) corresponds to 3000 kWh per apartment for the standard alternative (or 2125 and 1800 for base and best practices respectively). In the Base and Best Practices alternatives, 50% electricity is produced by photovoltaic solar panels and a 40% reduction in water consumption is considered. Lastly, solar domestic hot water systems provide 40% of the related energy requirements.

Public spaces are associated with the buildings. Two alternatives are envisaged: Standard Public Spaces and Base Public Spaces, with the following main characteristics:

The radar chart obtained by the ARIADNE software can be used to compare these alternatives (cf. figure below).

The environmental impacts of the (Base) project are considerably reduced in comparison to the Standard reference, except for damages due to ecotoxicity and health damage, which is linked to wood-fuelled heating. However, the inventory used for the wood-fuelled boiler is taken from the Ecoinvent 1.2 base, which gives an average for all existing boilers in Switzerland, and does not therefore correspond to a recent

Table 5.11 Life cycle assessment hypotheses for public spaces.

	Standard	Base and Best Practices
Electricity consumption for lighting streets/square/quay	30 kWh/(m².yr) (incandescent)	6 kWh/(m².yr) (sodium)
Electricity consumption for lighting green spaces	15 kWh/(m².yr)	3 kWh/(m².yr)
Waterproofing of streets	95%	95%
Waterproofing of quays & west paving of block C	85%	40%
Waterproofing of concrete paving	85%	85%
Waterproofing of green areas	25%	25%
Percentage of rainwater discharged into reservoir systems	10%	90%
Watering	Water supply	Recovered water

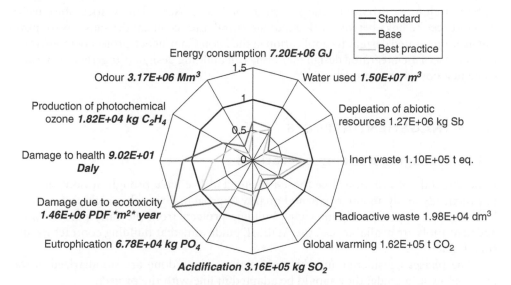

Figure 5.35 Life cycle assessment comparing 3 alternatives (ARIADNE software).

boiler. A specific inventory of boilers used in Lyon would improve the precision of this analysis. Given the good environmental practices chosen, the influence of public areas is very low in the neighbourhood, which would not be the case if conventional practices had been selected: their contribution could have reached 32% of primary energy consumption in the Best Practices alternative.

Lastly, because climate change is a major issue in the Concerto project, it is worth indicating the average emissions of public areas and buildings (cf. table below).

Thanks to their high thermal performance and significant integration of renewable energy, according to the calculation the buildings emit 64% less CO_2 than standard buildings. Public areas' emissions are reduced by 82%, in particular thanks to optimal

Table 5.12 GWP of buildings and public spaces (kg CO_2 eq./m^2/yr).

	Standard	Base	Best practices
Buildings	21.9	7.9	5.6
Public Spaces	2.2	0.4	0.4

lighting. One of the innovations of the Lyon Confluence neighbourhood was to set environmental performance targets, in terms of greenhouse gas emissions (7 kg eq. CO_2/m^2 per year for housing and 5 kg eq. CO_2/m^2 for tertiary) and generation of radioactive waste (2 g/m^2 per year). Integrating this kind of target into a programme is a significant step forward that would be worth repeating.

Regarding the methodological aspects, the aim of the software tools developed is to provide a design and decision-making aid for developers of real estate projects (property developers, municipalities, groups of investors, etc.), designers and producers (architects, urban designers, contractors), and users. This work's objective is to reinforce the links between urban, architectural and technical design so as to move towards integrated design. Its application in the Lyon Confluence project demonstrates the interest of an integrated design that brings all building designers together at an early stage in a common approach.

5.4 MANAGEMENT ASPECTS

5.4.1 Managing building operations

Experimental constructions like those presented above have brought lessons on how to practically apply an environmental quality approach. To reach a high performance level requires appropriately managing building projects from the early phases. Management tools are available, e.g. the ADEME guide aimed at building contractors and stakeholders[476].

The phases considered and the main actions to be done are summarized in the table below as a guide: they should be adapted in line with the project.

Learning to practise environmental quality in buildings usually leads to an awareness that some actions have not been carried out early enough (e.g. aid in choosing, advice before making a decision). For example, it is too late to change the orientation of a building in the basic preliminary design phase (it needs to be done at sketch stage). The summary table above provides some pointers for organizing environmental quality management in a complex process that involves numerous actors and is different for each operation: there are no ready-made procedures that can be applied systematically to all building projects. The process is an iterative one that must be adapted to each case and improved upon.

---

[476] Alain Bornarel et al., Qualité environnementale des bâtiments, manuel à l'usage de la maîtrise d'ouvrage et des acteurs du bâtiment, ADEME, Valbonne, April 2002

Table 5.13 Integrating environmental quality in the different project phases.

Phases	Actions and objectives
Feasibility of the operation	Evaluate needs, size a project adapted to these needs, estimate the investment and operation costs, (integrate energy and water savings)
Site	Choose a site, minimize environmental impacts (transport resulting from choice of site, presence of public transport, waste collection, exposure to sun, availability of energy – gas, district heating, etc.), minimize nuisance linked to the site (noise, pollution sources, etc.)
Choice of technical assistance	Seek out skills to put together a programme, monitor the building's design and construction and possibly monitor for one year after completion, recommended skills: lighting, acoustic and thermal calculations (preferably simulation) life cycle assessment
Programme	Define objectives and establish priorities through a participative approach involving all stakeholders, oversee construction costs by including optimization of life cycle cost (including use stage), integrate environmental quality targets in the programme, for example: – primary energy consumption (for heating, hot water, ventilation and lighting) $<50\,kWh/m^2/yr$, – low-flow sanitary equipment and equipment to sort activity waste, – minimal daylight factor >2 in main rooms, – no. of days per year where temperature exceeds $27°C$ <5 in main rooms for a typical climate year, – air renewal rate >0.6 air changes/hour in main rooms
Estimation of costs	Refine estimated construction cost, preferably taking a life cycle cost approach (including usage, maintenance, possibly "external" costs (environmental and social)
Design competition and choice of winner	Require skills from building project teams, in particular: lighting, acoustic and thermal calculations (preferably by simulation), life cycle assessment, pre-select teams taking into account their references (performance of previous constructions) and their note of intention Set up a jury and a technical commission that takes environmental criteria on board in their appraisal, set out selection criteria in a transparent manner, including explicit environmental criteria Select the winning team after analyzing "enhanced preliminary sketch": preliminary sketch accompanied by a note describing the main technical choices so that the most significant performances can be quantified (energy and water consumption, daylight factors, acoustic insulation, summer comfort, environmental impacts, life cycle cost), enhanced preliminary sketches constitute an intermediate level between the preliminary sketch and the basic preliminary design
Sketch	Define the general part of the construction (morphology), fit the building into the site, anticipate outside equipment (compost or waste sorting, bicycle garage, etc.), optimize ground waterproofing to manage rainwater, organize main functions, make key technical choices, respect economic constraints Optimize environmental quality by taking advantage of building's exposure, compactness, balance between glazed and opaque surfaces depending on direction of surfaces and spaces, reduce consumption of matter (e.g. lightweight surfaces in zones that do not require thermal mass): the tools described in chapter 3 can be used from this stage to refine and validate a sketch
Basic Preliminary Design	Describe the project with plans (1/200, possibly details at 1/100) showing each premises (including technical areas, circulation, etc.), justify technical measures, calculate a provisional estimate of the provisional cost of the work, Evaluate more precisely, using the tools described above, energy and water consumption, comfort levels (thermal, visual and acoustic), environmental impacts, life cycle cost, refine some parameters (insulation thickness, materials, glazing quality, etc.) to optimize environmental quality

(Continued)

Table 5.13 Continued.

Phases	Actions and objectives
Detailed Preliminary Design	Describe the project in a detailed manner (plans at 1/100, details at 1/50), write up precise technical notes (thermal, acoustic, lighting simulation), optimize equipment, e.g. ventilation (air flow ensuring satisfactory air quality, heat recovery, possibly night-time ventilation to cool in summer, hygro-adjustable or time-set ventilation, etc.), sanitary equipment (reduced flow), heating equipment (e.g. condensing, high-efficiency, or low NO_x emission boilers, solar systems, control, emitters), lighting systems (low-energy lamps and efficient light fittings, control), compare different energies to reduce environmental impact and costs (gas, wood, district heating, etc.), provide final estimate of provisional cost of the work
Tendering Package	Draw up the Special Technical Terms and Conditions by lot, preparing the order for contractors, and including detailed plans (1/50 or 1/20, in particular to deal with thermal bridges and air-tightness of the envelope), technical specifications (e.g. choice (of glazing and joinery) and block diagram schemes explaining how systems function
	Size installations (heating, ventilation, hot water, air conditioning), avoid over- sizing which can reduce performances (e.g. efficiency of boilers or air-conditioning units)
	Choose, for the same functionality, materials and Components (insulation, masonry, coatings, joinery, etc.) offering the highest environmental quality possible (cf. life cycle assessments, environmental and health declaration sheets, FSC wood certification), consider recycled, renewable or re-used materials that are easy to recycle, take into account impacts from cleaning products (especially when choosing coatings)
	Motivate partner companies (e.g. sub-contractors) by informing them of the quality approach and objectives, especially the "green worksite" approach (waste sorting, limited nuisances – noise, dust, liquid waste, etc.)
Assistance to building contracts	Select contractors based on responses to tendering package, taking into account environmental aspects in the responses (commitment to green worksite charter, choice of products, techniques proposed such as using recycled materials for foundations, reduction of thermal bridges, etc.), and possibly company references, oversee additional costs of eco-techniques reassuring companies about manufacturers' guarantees, facility of implementation and any references
Worksite	Prepare the worksite to limit nuisances to local residents (select access areas to manage traffic, limit noise and dust), reduce risks of accidents and pollution (collect used oil, polluting waste), anticipate storage zones to sort waste and for backfill,
	Inform stakeholders about environmental quality objectives at worksite meetings, inform local residents,
	Ensure that worksite operations conform to green worksite charter commitments (the presence of an environment monitor is required on-site), ensure that workers are trained (efficiency of sorting waste, appropriate pictograms, air-tightness), identify waste recuperation channels and control relationships between contractors
	Verify some technical parameters (reduction of thermal bridges, insulation thickness, air permeability, etc.)
Acceptance and follow-up	Check whether the construction conforms with initial programme and special terms & conditions (e.g. quality of windows, sanitary equipment, lighting systems, etc.), verify correct operation of the building, its components (windows, solar protection, etc.) and its equipment (heating, hot water, ventilation, lighting), carry out tests (blower door for air tightness, infra-red thermography)

(Continued)

Table 5.13 Continued.

Phases	Actions and objectives
	Send future occupants notices, plans and information on managing and maintaining the building, inform them about environmental quality objectives and their role in the building's actual performance (temperature settings, ventilation, water and energy consumption, waste sorting, building maintenance, etc.) Monitor the building's performance during the first year (full guarantee of perfect completion for the year): note energy consumption (measurements may be perturbed in year 1 as some components dry out and controls are refined, and measuring during the second year is recommended), measure water consumption, temperatures, lighting levels, noise levels, and other more qualitative parameters with a satisfaction survey of occupants
Interior and exterior furnishings	Choose interior furnishings and equipment not included in the actual construction (light fittings, domestic and office appliances, etc.), taking into account water and energy savings (energy label for domestic appliances, priority for classes A & B), potential emissions from these appliances (e.g. formaldehyde or other VOCs linked to some glues or varnishes used on furniture), reduced packaging, easier maintenance and impacts generated at this stage, use of recycled or renewable materials, re-used or recyclable products Install appliances that generate electro-magnetic flows so as to limit human exposure (avoid placing this kind of appliance close to beds, work stations, etc.) Choose plantations for green areas so as to limit impacts from maintenance and the need for watering
Building management	Ensure occupants' comfort and the building's smooth running while limiting environmental impacts, especially: – adjust temperatures over time so as to avoid over-heating and reduce heating during unoccupied periods, – reduce ventilation during unoccupied winter periods, increase it in summer if night cooling is necessary, – inform occupants of their role in managing the building (managing lighting, water consumption, sorting waste, car-pooling or use of public transport, cycling, etc.), – maintain the building so as to increase the lifespan of components (e.g. regularly paint woodwork, maintain equipment for heating, hot water, ventilation, etc.), – maintain the building so as to limit health risks (check domestic hot water installation, clean filters and vents, clean premises using products with low environmental impact, etc.), – measure energy and water consumption to detect excesses (fault in the boiler or heating control, water leak, etc.), – link up the building's equipment with municipal management (waste sorting), – choose consumables with low environmental impact (e.g. recycled paper)
Retrofit	Retrofit operations follow a fairly similar process to new construction projects: at the feasibility phase, an audit can evaluate requirements by consulting inhabitants and making technical assessments (e.g. energy audit); resources that can be mobilized are evaluated (common equity, loan repaid in line with rents, grants, etc.). It is not common for this operation to be open to competition since the building's architecture is rarely called into question. An architect is however involved in designing the façades. Analysis tools (for heat, lighting, acoustics, life cycle) can be used to compare alternatives and optimize choices, from basic preliminary design to special terms and conditions. Subsequent phases are similar to those of a new construction (company contracts, worksite, acceptance, management)
Deconstruction	Ensure that the building's deconstruction is necessary (possibly compare with a retrofit scenario) Prefer deconstruction to demolition, i.e. dismantling all components that can be separated, e.g. windows, possible re-use of components, sorting and recycling of waste Manage the worksite taking a green approach (cf. worksite phase), with even more emphasis on waste management (volume and cost).

5.4.2   Co-housing approach

The above description of environmental management shows the importance of making early choices, right from the first stages of a project (choice of terrain, sketch). In the case of a residential building project, a property developer generally makes the decisions during these phases. The current practice is to limit the cost of the building so as to make it easier to sell the apartments or houses.

In this set-up, it is very difficult to integrate environmental quality aspects: even though some energy- and water-saving techniques are cost efficient overall, property developers are often unlikely to agree to adding initial costs to building work and incurring the risk of not being able to sell a product that will then be more expensive than other products on the market. The result is a degree of standardization in housing, aimed at pleasing the majority. Yet some purchasers seek greater environmental quality. They can buy a plot and have their own house built, but another approach is possible: co-housing, which allows people to take a more sustainable development approach by applying it on a neighbourhood scale and pooling some of the costs (i.e. studies, purchase of materials in greater quantity, etc.).

The co-housing concept involves purchasers agreeing to a set of basic environmental quality objectives and having a neighbourhood built comprising houses, roads, collective equipment and a community centre. The centre provides a range of surfaces, e.g. a large room for community meals or meetings, guest bedrooms, an office, a games room for children, a laundry. The streets close to the houses may be pedestrian, helping to keep down noise levels and providing safe areas for children to play. Working together on devising a habitat results in improvements to the project and economies of scale. The expression "co-housing" first emerged in Denmark[477,478], but projects had already been carried out in France in the 1980s[479].

The figure 5.36 below shows one of these neighbourhoods, located in Golden near Denver (Colorado, USA). It comprises a set of thirty houses. The garages are placed on the outside so that the houses can be laid out in pedestrian walkways. Along with the community house (400 m^2), gardens and play areas are included. The architectural style is inspired by the Pueblo Indians as a reference to their community culture and the region's history. The houses range in size from 80 m^2 to 350 m^2, depending on the number of inhabitants and rooms. The purchasers put together a list of quality criteria to guide the design: easy maintenance, low energy consumption through passive solar architecture, low cost, light and space, balance between private and public spaces, and environmentally friendly materials. Increased insulation saves energy, along with low-emissivity windows and controlled ventilation. According to the local evaluation procedure (no national regulation exists on heating buildings), the houses are 30% to 40% more economical than on average. The basic design can be supplemented by "green" options: recycled carpets, recycled cellulose insulation, low-VOC paint.

---

[477] Chris Hanson, *The cohousing handbook*, Hartley and Marks Publishers, Point Roberts, 1996
[478] Kathryn McCamant and Charles Durett, *Cohousing, a contemporary approach to building ourselves*, Ten speed press, Berkeley, 1994
[479] MGHA (Mouvement pour l'habitat groupé autogéré), Habitats Autogérés, Editions Alternatives/Syros, Paris, June 1983

Figure 5.36 Ecological neighbourhood in Golden (co-housing).

Figure 5.37 Community centre in a co-housing neighbourhood (Denver, USA).

Each family owns one house, with a shared community centre comprising a large dining room and a community kitchen, a television room, a laundry, play areas for the children, etc. Figure 5.37 above shows a community centre in another co-housing neighbourhood in Denver.

Organizing this kind of project requires investing time and involves different stages, e.g. creating an informal group; setting up a first legal body for drawing up a contract with a property developer; a feasibility study from the property developer; setting up a second body grouping the preceding body and the property developer to apply for a bank loan, buy the land and build; the sketch, detailed project design and building permit request; demands for estimates and possible adaptation of the project (suitability of the budget); attributing the contracts; monitoring the worksite; accepting the work; managing the buildings.

Similar experiences sometimes occur today with the eco-village concept. The European e-co-housing project grouped different partners with the aim of developing tools to integrate a participative approach and environment protection into neighbourhood design[480].

5.5 PERFORMANCE PROSPECTS

5.5.1 The "passive house" standard

Numerous experiments in the energy domain have identified good practice in design and production, especially:

- applying increased insulation to envelopes;
- using high-performance windows, preferably south-facing;
- reducing thermal bridges and controlling air infiltrations;
- recovering energy by balanced ventilation;
- using solar water heaters and low-flow sanitary equipment;
- managing electricity demand for lighting and domestic appliances.

Buildings designed and constructed in line with these principles have shown that high performances can be achieved at a reasonable cost. The table below gives examples of energy balance sheets based on the "Passive house" label[481] and a standard respecting current heating regulations[482] for gas heating and electric heating. A factor of 2.58 is considered for the equivalent of one electric kWh of primary energy, and the average efficiency of the gas boiler is 86%. Auxiliary equipment corresponds to single-flow mechanical ventilation in the RT 2005 case, along with a circulator in the

---

[480]Bruno Peuportier, Towards sustainable neighbourhoods, the eco-housing project, IV International Conference "Climate change – energy awareness – energy efficiency", Visegrad, June 2005

[481]Wolfgang Feist, *Gestaltungsgrundlagen Passivhaüser*, Verlag das Beispiel, 2001

[482]Ministry for employment, social cohesion and housing, decree of 24 May 2006 on thermal characteristics of new buildings and new parts of buildings, Official Journal, 25 May 2006

Table 5.14 Examples of primary energy balance sheets regulation (ZONE H1[483]) and "passive house" label.

Energy ratios in kWh/m²/yr	RT 2005 regulation		Passive House label + solar water heater
	Gas	Electricity	
Heating load	60	62	13
Primary energy for heating	70	159	15
Primary energy for domestic hot water	30	68	10 (low-flow)
Primary energy for lighting	15	15	7 (energy-saving)
Auxiliary equipment	15	8	10
Total	130	250	40

case of gas heating, and balanced ventilation in the passive house (with electronically adjusted ventilators).

In France, a residential building was constructed as part of the European project CEPHEUS with the aim of promoting passive construction – the Salvatierra building in Rennes. Relatively simple technologies were used, resulting in reasonable building costs. In addition, the very high performance of the envelope keeps the heating load at such a low level that heating can be provided using the ventilation air, thus bringing down the construction cost since the radiators and water loop are no longer necessary. The life cycle balance sheet for this type of building is clearly of interest, since the materials used are not very sophisticated and the flows are very low during the use stage. Nevertheless, the clay and straw mortar adversely affected the building's air-tightness, and the 15 kWh/m² threshold required by the label was not achieved: the annual heating load is slightly over 20 kWh/m². However, numerous other projects are underway or completed, and a general application of low-energy levels (total annual consumption of primary energy under 65 kWh/m² in the same climate zone) has been suggested for thermal regulations by 2012.

New techniques for integrating solar components in buildings, e.g. thin films in photovoltaic cells, are devised to ultimately produce more energy than the building consumes. These kinds of building are thus overall energy producers, but they usually need to be connected up to the grid to ensure that energy supply matches demand at the right time (except for remote buildings, where energy is stored in batteries).

5.5.2   Energy-producing buildings

Beyond the "passive house" standard, it is possible to achieve "zero energy", or even to produce energy in a building. This starts with reducing consumption, e.g. by applying the passive house standard, then using techniques for producing energy that make it possible to either compensate the use of energy from the network (especially electricity) by supplying energy (at which point zero energy is achieved), or to ensure net energy production.

---

[483]The "H1" climate zone of the French thermal regulation corresponds to the coldest regions: the northeast of the country (cf. www.rt-batiment.fr)

The ZEN house, close to the town of Chambéry, was designed on this principle by the company CYTHELIA[484], cf. illustration below. The solid wood envelope associates increased insulation (external cork insulation, low-emissivity argon-filled triple glazing) with a bioclimatic design, which reduces the heating load. This load is covered by a heat pump using exhaust air as a heat source, coupled with an air/water heat pump for domestic hot water. The total annual consumption of electricity (heating, hot water, ventilation, lighting and other uses) is evaluated at 45 kWh/m^2. The production, which is assured by photovoltaic panels (CIS thin film) covering the entire roof (i.e. 13.5 kWc), should be higher, and measurements are currently being done on this building.

Figure 5.38 Zero energy house in Chambéry (Cythelia).

In August 2008, the house's consumption rate, which had been occupied by Cythelia since January, reached 4800 kWh for a solar production of 8400 kWh.

The following example is an energy-producing house in the USA. The Sawyer house was designed with a bioclimatic approach (Trombe wall to the south, few windows to the north, balanced ventilation, very good thermal insulation). The heating load is thus reduced, and is almost completely covered by direct solar floor heating, fed by

---

[484]Alain Ricaud, Cythelia Expertise et Conseil, Presentation pack on La maison ZEN, Montagnole, August 2007

Figure 5.39 Zero energy school in Limeil Brevannes.

vacuum tube collectors (which can also be used to produce domestic hot water) with hot water storage. Additional heat can be supplied by a heat pump, but the inhabitants say they have not yet had to use it. A 11 kWp photovoltaic system and a small wind turbine provide electricity, including for an electric car, and even generate surplus electricity injected into the grid and sold to the distributor. The house is therefore energy-producing overall.

The Jean-Louis Marquèze School (5 nursery classes and 7 primary classes) was opened in 2007 in Limeil Brevannes, cf. illustration below. It was the first school to be built in France with a positive energy target. The means used to reach the objective were: bioclimatic design, reinforced insulation (21 cm in the walls, triple glazing), balanced ventilation adjusted according to occupancy (CO_2 content), heat pump on the water table and solar integration (30 m^2 of solar thermal collectors and 800 m^2 of photovoltaic modules on the roof and façade). The energy balance may confirm calculations, with the main uncertainties being the envelope's tightness. Other similar projects are underway, e.g. in Pantin.

More extensive use of energy-producing buildings by 2020 was recently envisaged in France during the "Grenelle de l'Environnement" consultation meetings. Research is being done to evaluate the impact of this kind of building on the electricity grid.

5.5.3 Passive retrofit

The examples above show that it is possible to attain very high performances with new constructions, but the challenge – and difficulty – lies mainly in existing buildings. The experiments described above in the social housing sector show that performances can come close to those of passive buildings if techniques like external insulation and balanced ventilation can be installed, but this is not always possible for architectural reasons (e.g. protected façades so that insulation must be on the inside) or technical ones (presence of load-bearing partition making it difficult to pass air ducts through, roof position unsuitable for integrating solar components, lack of place for installing a solar hot water tank for collective heating, etc.).

Technical solutions have therefore been developed to solve these problems, e.g. vacuum insulation can cut down the loss of habitable surface area by reducing the thickness of insulation required, but it remains expensive; individual balanced ventilation can avoid having to pass air ducts from one room to another, but the correct distribution of fresh air must be verified.

In all cases, retrofit requires considerable investment[485], skills and new organization methods: e.g. air infiltration work entails coordination between different professions (rough work, light work, electricity, etc.). Regulations obliging overall performance only apply to extensive retrofits (i.e. the cost of the work must be more than 25% of a reference building cost), and concerns very few operations. Regulations on individual components could be more strict, and for example oblige the use of rare gases in double glazing: the supplementary cost of a few euro is quickly covered and it is a shame to continue installing air-filled windows in residential and tertiary buildings.

During this period of tensions over resources, in particular energy, it is once again time to study new technical and financial devices ("green" loans), regulations (national and municipal) and organization tools. Germany has set up a system of subsidised loans as part of a programme to retrofit buildings. Loans amounted to one and a half billion euro in 2006, 50% more than anticipated, and 256,000 houses were renovated. A similar initiative has recently been launched in France.

5.5.4 Towards a cyclic economy

In addition to energy aspects, the depletion of some raw materials is starting to take effect, in particular metals. The building sector, which consumes high amounts of materials, needs to adapt: using up raw materials and transforming them into waste is not sustainable. Projects are starting to emerge whereby the co-production of some industries (that used to be considered as waste) feeds into other processes. Although it is still difficult to put into place, this kind of cyclic economy system will no doubt be a key part of the future.

---

[485] From 6 to 7 billion euro for 450,000 houses and 15 million m² tertiary buildings per year, i.e. the equivalent of building 80,000 social houses, cf. Olivier Sidler, Réglementation énergétique dans les bâtiments antérieurs à 1975, www.negawatt.org

Taking "grey"[486] energy into account, which is set to come under regulations by 2020, is a first step towards promoting the efforts made in terms of environmental quality over a building's life cycle.

5.6   CONCLUSIONS OF CHAPTER 5

The numerous experiments carried out, in particular in the energy field, have made it possible to identify good practice and make progress with technologies thanks to feedback on their implementation and users' acceptance. Several general conclusions can be drawn:

- regarding performance, very low energy requirements can be attained, and it is even possible to produce more energy than that consumed in a building; life cycle assessments should be implemented to supplement these energy evaluations and so reduce environmental impacts; it is best to integrate environmental performance into construction programmes;
- regarding technologies, products are now much more reliable, additional investment costs can disappear if an innovative technology becomes standard (cf. example of low-emissivity glazing), research is being done to make these innovations easier to implement (e.g. the integration of balanced ventilation into existing buildings);
- regarding project management, work remains to be done to move from a suboptimal set-up (where each actor only optimizes one stage or technical domain) to a decision process integrating the entire life cycle of a construction and all professional fields (architecture, finance, heat, acoustics, lighting, environment, etc.);
- an important aspect of management concerns quality checking: a programme should include performance requirements with measurable indicators, and at least annual water and energy consumption. The lack of precision on these performance levels and on monitoring operations limits the usual "HEQ" approach; the European directive 2002/91/EC[487] on the energy performance of buildings should change this situation with the obligatory display of energy consumption in place since 4 January 2006;
- involving users can open up new possibilities for choosing design, and energy-aware behaviour is an essential part of a building's performance.

---

[486] Energy consumed to manufacture, install, maintain and deal with the end of life of materials and components (named "grey" because it does not show up on energy bills, which only concern the use stages of a building).

[487] Directive 2002/91/EC of the European Parliament and Council of 16 December 2002 on energy performance of buildings, Official Journal of the European Communities, 4 January 2003

Conclusions

Solar architecture, passive solar, bioclimatic design, high environmental quality, low energy, positive energy, zero emission: the vocabulary changes in line with the trend, and each new trend brings a new idea. We need to capitalize on knowledge and develop performance in the long term.

Improving performance requires integrating multiple quality aspects. We should foster exchanges between the different partners (architects, engineers, environmentalists, doctors, ergonomists, construction economists, etc.) and encourage the users of buildings to participate.

Evaluations of environmental impacts supplement analyses made to date, for example on thermal and acoustic aspects. They can involve doing a life cycle assessment, as long as the functional unit considered is clearly defined, and clear boundaries are set for the system studied (building, site, occupants) in line with the study's objective. Numerous uncertainties remain regarding the data used and environmental quality indicators, and it is important to integrate the progress of international knowledge on the subject. Different European projects such as REGENER[488] and ECO-HOUSING[489] have gathered several teams to define an initial methodological framework. The European network PRESCO[490] has produced a comparison of the different approaches and work is still in progress on two projects involving a dozen partners: ENSLIC Building and LORE LCA.

Industrials provide information on the environmental quality of building products in the form of databases[491]. Hopefully, these bases will become more precise and transparent, like the Ecoinvent database, which boosts quality by including several hundred substances in its inventories along with several thousand pages of documentation. Information for users of buildings is also making progress: in particular the

---

[488]REGENER final reports, C.E.C. DG XII contract No. RENA CT94-0033, January 1997, 563 p

[489]ECO-HOUSING, Environmental co-housing in Europe, European project number: NNE5/2001/551, publishable final report, January 2006, 29p

[490]Bruno Peuportier, Katrien Putzzys et al., *Inter-comparison and benchmarking of LCA-based environmental assessment and design tools*, final report, February 2005, http://www.etn-presco.net/

[491]ECOINVENT database in Switzerland (including numerous European data): www.ecoinvent.ch, INIES in France: www.inies.fr, www.greenbooklive.com in Great Britain, www.rts.fi/english.htm in Finland.

European directive 2002/91/EC of December 2002[492] making it obligatory to display energy consumption has been transposed in France, and the energy label has become standard, although it will take several years to train specialists.

Life cycle assessments reveal how important energy is in a building's environmental balance. They can be used to guide development and assess the benefits of using innovative technologies. In addition, they help professionals, architects and engineers make decisions to improve projects' environmental quality, in particular during the design phase.

These new measures are supported by technologies that take environmental aspects into account, making it easier to save on resources, recycle materials and minimize polluting emissions.

These developments in building sector practices can make it easier to respect some international commitments, in particular regarding climate change. We hope that this publication will contribute to extending the array of methodological and technological tools available to designers.

---

[492] Directive 2002/91/EC of the European Parliament and of the Council of 16 December 2002 on the energy performance of buildings, Official Journal of the European Communities, 4 January 2003

Bibliography

ENVIRONMENT

ADEME, *www.ademe.fr*

AFNOR, *norme X 30-300, Analyse du cycle de vie*, March 1994

AFNOR, *norme NF P01-010, Qualité environnementale des produits de construction – Déclaration environnementale et sanitaire des produits de construction*, December 2004

AFNOR, *norme NF P01-020-1, Bâtiment – Qualité environnementale des bâtiments – Partie 1: cadre méthodologique pour la description et la caractérisation des performances environnementales et sanitaires des bâtiments*, March 2005

R. Barbault, *Écologie générale*, Ed. Masson, 1990

Daniel Bernstein, Jean-Pierre Champetier, Loïc Hamayon, Ljubica Mudri, Jean-Pierre Traisnel and Thierry Vidal, *Traité de la construction durable*, Ed. Le Moniteur, Paris, 2007

Alain Bornarel et al., *Qualité environnementale des bâtiments, manuel à l'usage de la maîtrise d'ouvrage et des acteurs du bâtiment*, ADEME, Valbonne, April 2002

Catherine Charlot-Valdieu and Philippe Outrequin, *Analyse environnementale d'un quartier urbain dans une perspective de développement durable*, Cahier CSTB n° 3236, July–August 2000

Jean-Luc Chevalier and Jean-François Le Téno, *Analyse de l'impact environnemental du cycle de vie des produits de construction*, rapport final ADEME, CSTB Valbonne, September 1994

CITEPA (Centre Interprofessionnel Technique d'Etudes sur la Pollution Atmosphérique), *Inventaire des émissions de polluants atmosphériques en France – séries sectorielles et analyses étendues*, Feb. 2008, www.citepa.org

CITEPA, *Inventaire des émissions de polluants atmosphériques en France au titre de la convention sur la pollution atmosphérique transfrontalière à longue distance et de la directive européenne relative aux plafonds d'émission nationaux*, Dec. 2007

Marcel Deneux, *L'évaluation de l'ampleur des changements climatiques, de leurs causes et de leur impact prévisible sur la géographie de la France à l'horizon 2025, 2050 et 2100*, Office parlementaire des choix scientifiques et technologiques, January 2002

Frédéric Denhez, *Les pollutions invisibles*, Ed. Delachaux et Niestlé, 2005

Suzanne and Pierre Déoux, *L'écologie c'est la santé*, Ed. Frison Roche, 1994

Suzanne and Pierre Déoux, *Le guide de l'habitat sain*, Ed. Medieco, 2002

Benjamin Dessus, Hervé Le Treut and Bernard Laponche, *Effet de serre, n'oublions pas le méthane*, La Recherche, No. 417, March 2008

Ecoinvent database, www.ecoinvent.ch

Ed. Le Vent du Chemin, *Manuel d'écologie urbaine et domestique*, St Denis, 1992

Christian Elichegaray, *Retombées des polluants acides: vers des normes draconiennes au 21e siècle*, Agence pour la qualité de l'air, 1991

EUSES the European Union System for the Evaluation of Substances. National Institute of Public Health and the Environment (RIVM), The Netherlands; Available from the European Chemicals Bureau (EC/JRC), Ispra, Italy, https://ec.europa.eu/jrc/en/scientific-tool/european-union-system-evaluation-substances, *EUSES 2.1 Background report*, July 2008

Claude François and Bruno Hilaire, *Guide pour les économies d'eau*, Cahiers du CSTB No. 422, September 2001

R. Friedrich, Ari Rabl and Joe Spadaro 2001, *Quantifying the Costs of Air Pollution: – The ExternE Project of the European Commission – Quantification des coûts de la pollution atmosphérique: le projet ExternE de la Commission européenne*, Atmospheric Pollution, Special bilingual Issue, pp. 77–104, Dec. 2001

Rolf Frischknecht et al., *Oekoinventare von Energiesystemen, 3. Auflage*, ETH Zürich/PSI Villigen, 1996

Rolf Frischknecht, Jungbluth N., Althaus H.-J., Doka G., Heck T., Hellweg S., Hischier R., Nemecek T., Rebitzer G., Spielmann M., *Overview and Methodology, ecoinvent report No. 1*, Swiss Centre for Life Cycle Inventories, Dübendorf, Switzerland, 2004

U. Fritsche et al., *Total Emission Model for Integrated Systems (TEMIS), Version 2.0 Manual*, Öko-Institut Darmstadt, 1994

U. Fritsche et al., *Gesamt-Emissions-Model Integrierter Systeme (GEMIS), Version 2.1 Erweiterer Endbericht*, Öko-Institut Darmstadt, 1994

Javier Garcia et al., *Les indices de qualité de l'air*, Presses de l'Ecole des Mines de Paris, September 2001

Michel Gerber and ADEME, *Cahier des charges à l'usage des concepteurs*, Languedoc-Roussillon, Oct. 1991

Mark Goedkoop, *Weighting method for environmental effects that damage ecosystems or human health on a European scale*, NOVEM, Utrecht, 1995

Mark Goedkoop and R. Spriemsma, *The Eco-Indicator 99, A damage oriented method for life cycle impact assessment, methodology report, methodology annex, manual for designers*, April 2000

Guinée J. B., (final editor), Gorrée M., Heijungs R., Huppes G., Kleijn R., de Koning A., van Oers L., Wegener Sleeswijk A., Suh S., Udo de Haes H. A., de Bruijn H., van Duin R., Huijbregts M. A. J., Lindeijer E., Roorda A. A. H., Weidema B. P.: *Life cycle assessment; An operational guide to the ISO standards;* Ministry of Housing, Spatial Planning and Environment (VROM) and Centre of Environmental Science (CML), Den Haag and Leiden, The Netherlands, 2001, 704 p.

R. Heijungs, *Environmental life cycle assessment of products, Centre of environmental science (CML)*, Leiden, 1992, 96 p.

Olav Hohmeyer and Michael Gärtner, *The cost of climate change: a rough estimate of orders of magnitude*, The yearbook of renewable energies, EUROSOLAR/CEE, 1994

Jean-Charles Hourcade, *Traitement de l'innovation et évaluation des coûts à long terme de la réduction des émissions de CO_2*, Revue de l'Energie, No. 427, January 1991

INIES database, www.inies.fr

IPCC, Forster, P.M. (2007) Changes in Atmospheric Constituents and in Radiative Forcing, in: Solomon, S., D. Qin, M. Manning, Z. Chen, M. Marquis, K.B. Averyt, M. Tignor and H.L. Miller (ed.), *Climate Change 2007: The Physical Science Basis. Contribution of Working Group I to the Fourth Assessment Report of the Intergovernmental Panel on Climate Change*, Cambridge University Press, Cambridge, United Kingdom and New York, NY, USA, 2007

IPCC, Houghton J. T., Ding Y., Griggs D. J., Noguer M., van der Linden P. J. and Xiaosu D.: *Climate Change 2001: The Scientific Basis*, IPCC, Intergovernmental Panel on Climate Change, Cambridge University Press, The Edinburgh Building Shaftesbury Road, Cambridge, UK, ISBN-13: 9780521014953, July 2001, 892 p.

IPCC, Scientific assessment working group of IPCC, *Radiative forcing of climate change*, World meteorological organization and United Nations Environment Programme, 1994, 28 p.

Olivier Jolliet, Myriam Saadé and Pierre Crettaz, *Analyse du cycle de vie, comprendre et réaliser un écobilan*, Presses Polytechniques et Universitaires Romandes, Lausanne, 2005

Olivier Jolliet et al., *IMPACT 2002+, A new life cycle impact assessment methodology*, International Journal of Life Cycle Assessment, vol. 8, No. 6, pp. 324–330, 2003

Philippe Kerjolis et al., *Code Permanent Environnement et Nuisances*, Ed. législatives, 2000

Niklaus Kohler et al. EPFL-LESO/IFIB (Karlsruhe University), *Energie- und Sofffluß-bilanzen von Gebäuden während ihrer Lebensdauer*, Ifib – Karslruhe, June 1994, 221 p.

Friedrich Kur, *L'habitat écologique, quels matériaux choisir ?*, Ed. Terre Vivante, 1999

Éric Labouze, *Bâtir avec l'environnement, enjeux écologiques et initiatives industrielles*, Ed. de l'Entrepreneur, Paris, 1993

Serge Lambert et al., *Manuel environnement à l'usage des industriels*, AFNOR, 1994

Amory and Hunter Lovins, Ernst Von Weizacker, *Facteur 4*, Ed. Terre Vivante, 1997

La Maison écologique, bimonthly magazine, www.lamaisonecologique.com

J. Mallevialle and T. Chambolle, *La qualité de l'eau*, La Recherche no. 221

Yves Martin, *L'effet de serre, rapport de l'académie des sciences No. 31*, November 1994

Jean Matricon, *Vive l'eau*, Ed. Gallimard, 2000

MGHA (Mouvement pour l'habitat groupé autogéré), *Habitats Autogérés*, Editions Alternatives/Syros, Paris, June 1983

Murray C. and Lopez, A, *The Global Burden of Disease*, WHO, World Bank and Harvard School of Public Health, Boston, 1996, 990 p.

OECD, *Towards sustainable development – environmental indicators*, March 1998

Bruno Peuportier, Niklaus Kohler and Chiel Boonstra, *European project REGENER, life cycle analysis of buildings*, 2nd International Conference " Buildings and the environment ", Paris, June 1997, pp. 33–40

Bruno Peuportier et al., *EU Project EASE (Education of Architects on Solar Energy and Environment)*, ALTENER final project report No. 4.1030/Z/98-340, European Commission, DG TREN, August 2000

Bruno Peuportier, Emil Popovici and Maxime Trocmé, *Analyse de cycle de vie à l'échelle du quartier*, Séminaire ADEQUA, Chambéry, October 2006

Bernd Polster, *Contribution à l'étude de l'impact environnemental des bâtiments par analyse du cycle de vie*, Doctorate Thesis, Ecole des Mines de Paris, 1995, 268 p.

Emil Popovici, *Contribution à l'analyse de cycle de vie des quartiers*, Doctoral Thesis, Ecole des Mines de Paris, 209 p, 2006

PRe Consultants, *SIMA-PRO* (Dutch software for life cycle analyses, used in the industry sector), 1993

François Ramade, *Écologie appliquée*, Ediscience, Paris, 1995

William Rees and Mathis Wackernagel, *Our ecological footprint: reducing human impact on the earth*, New Society Publishers, Gabriola Island, Canada, 1996

Anne Rialhe and Sylviane Nibel, *Quatre outils français d'analyse de la qualité environnementale des bâtiments*, Ed. Plan Urbanisme Construction et Architecture, 1999

Randall Thomas, *Environmental Design: an Introduction for Architects and Engineers*, E & FN Spon, London, 1996

THERMAL PERFORMANCE

Patrick Bacot, *Analyse modale des systèmes thermiques*, Doctoral Thesis, Paris VI University, 1984

Patrick Bacot, Alain Neveu and Jean Sicard, *Analyse modale des phénomènes thermiques en régime variable dans le bâtiment*, Revue Générale de Thermique, No. 267, Paris, 1984

Isabelle Blanc Sommereux and Gilles Lefebvre, *Simulation de bâtiment multizone par couplage de modèles modaux réduits*, CVC, No. 5, May 1989

Isabelle Blanc Sommereux and Bruno Peuportier, *A bioclimatic design aid based on multizone simulation*, ISES Conference, Denver, 1991

Dominique Blay, *Comportement et performance thermique d'un habitat bioclimatique à serre accolée*, Bâtiment-Energie, No. 45, 1986

Dominique Campana, François Neirac and Gabriel Watremez, *Elaboration d'un logiciel sur micro ordinateur pour l'aide à la conception des bâtiments en pays tropicaux secs*, rapport final REXCOOP, 1985

C. Carter, *A validation of the modal expansion method of modelling heat conduction in passive solar buildings*, Solar Energy, 23, No. 6, 1979

Joe A. Clarke, *Energy simulation in building design*, Adam Hilger Ltd, Bristol and Boston, 1985

Comité de Liaison Energies Renouvelables, www.cler.org

Duffie, Beckman, *Solar Engineering of Thermal Processes*, Solar Energy Laboratory, University of Wisconsin, Madison, 1980

Fanger P.O., *Thermal Comfort*, Mc Graw Hill, New York, 1970

Wolfgang Feist, *Gestaltungsgrundlagen Passivhaüser*, Verlag das Beispiel, 2001

B. Givoni, *L'homme, l'architecture et le climat*, ed. du Moniteur, Paris, 1978, 460 p.

Adolf Goetzberger et al., Transparent Insulation special issue, Solar Energy, Vol. 49, No. 5, pp. 331–449, November 1992

Gilles Lefebvre, *Analyse et réduction modale d'un modèle de comportement thermique de bâtiment*, Doctorate Thesis, University Paris VI, 1988

H. Lund, *Short Reference Years and Test Reference Years for EEC countries*, EEC Contract ESF-029-DK, 1985

Association La maison passive France, www.lamaisonpassive.fr

Jacques Michel, Patent ANVAR TROMBE MICHEL BF 7123778 (France 29/06/1971) and addition: Patent "Stockage thermique" MICHEL DIAMANT DURAFOUR 75-106-13

Alain Neveu, *Étude d'un code de calcul d'évolution thermique d'une enveloppe de bâtiment*, Doctorate Thesis, University Paris VI, 1984

Solar tools, www.outilssolaires.com

Bruno Peuportier and Isabelle Blanc-Sommereux, *Simulation tool with its expert interface for the thermal design of multizone buildings*, International Journal of Solar Energy, 1990, vol. 8, pp. 109–120

Bernd Polster, *Contribution à l'étude de l'impact environnemental des bâtiments par analyse du cycle de vie*, Doctorate Thesis, Ecole des Mines de Paris, 1995

Jean-Jacques Salgon and Alain Neveu, *Application of modal analysis to modelisation of thermal bridges in buildings*, Energy and buildings, October 1987

Thierry Salomon and Stéphane Bedel, *La maison des négawatts*, éd. Terre Vivante, Mens, July 1999

Solar systems, www.energies-renouvelables.org

Françoise Thellier, *L'Homme et de son environnement Thermique, Modélisation*, Accreditation to supervise reserach, Paul Sabatier University. Toulouse, 1999, 60 p.

Stéphane Thiers and Bruno Peuportier, *Modélisation thermique d'un échangeur air-sol pour le rafraîchissement de bâtiments*, Theme Day, SFT-IBPSA-France, Aix-les-Bains, April 2007

LIGHTING AND ELECTRICITY

AFE, *Les sources de lumière*, Société d'édition LUX, 1987

AFNOR, *Norme NF X 35-103, Principe d'ergonomie visuelle applicable à l'éclairage des lieux de travail*, 1990

Bouvier François, *Eclairage Naturel*, Techniques de l'Ingénieur, C 3315, February 1988

CSTB, Michel Perraudeau, *Distribution de la luminance du ciel, Études et Recherches*, Centre Scientifique et Technique du Bâtiment, Cahier 2305, livraison 295, December 1988

CSTB, Michel Perraudeau, *Mixte, logiciel d'éclairage mixte sur PC. Intégration des modules DISPO et CARPHOT*, CSTB, EN-ECL 94.4.2, Centre de Recherche de Nantes, Division Eclairage et Colorimétrie, 1994

DOE, *Advanced Lighting Guidelines*, US Department of Energy, California Energy Commission, Electric Power Research Institute, DOE/EE-0008, 1993

Fontoynont Marc R., *Prise en compte du rayonnement solaire dans l'éclairage naturel des locaux: méthodes et perspectives*, Doctoral Thesis, Ecole Nationale des Mines de Paris, 1987

Alain Guiavarch, *Etude de l'amélioration de la qualité environnementale du bâtiment par intégration de composants solaires*, Doctoral Thesis, Cergy-Pontoise University, November 2003

IES, *Lighting handbook, Application Volume*, Illuminating Engineering Society of North America, 1987

IES, *Recommended Practice for the Lumen Method of Daylight Calculations*, IES Report RP-23, Illuminating Engineering Society of North America, 1989

Marc La Toison, *Matériels et projets*, Techniques de l'ingénieur, C 3341, 1997

Marc La Toison, *Sources de lumière de l'éclairage électrique (détails techniques sur différents types de lampes)*, Techniques de l'ingénieur, Génie électrique/Utilisation de l'électricité II. D 5810, 1992

Marc La Toison, *Eclairage, données de base*, Techniques de l'ingénieur, C 3340, 1986

Ari Rabl and Jan Kreider, *Heating and Cooling of Buildings*, McGraw-Hill, 1994

Alain Ricaud, *Photopiles solaires*, Presses Polytechniques et Universitaires Romandes, 352 p, 1997

Olivier Sidler, *Connaissance et maîtrise des consommations des usages de l'électricité dans le secteur résidentiel*, May 2002, cf. www.enertech.fr

UTE, *Méthode simplifiée de prédétermination des éclairements dans les espaces clos et classification correspondantes des luminaires*, NF C 71–121, May 1993

Index